T0211438

IoT Solutions in Microsoft's Azure IoT Suite

Data Acquisition and Analysis in the Real World

Scott Klein

Apress®

IoT Solutions in Microsoft's Azure IoT Suite: Data Acquisition and Analysis in the Real World

Scott Klein
Redmond, Washington, USA

ISBN-13 (pbk): 978-1-4842-2142-6   ISBN-13 (electronic): 978-1-4842-2143-3
DOI 10.1007/978-1-4842-2143-3

Library of Congress Control Number: 2017939347

Copyright © 2017 by Scott Klein

This work is subject to copyright. All rights are reserved by the Publisher, whether the whole or part of the material is concerned, specifically the rights of translation, reprinting, reuse of illustrations, recitation, broadcasting, reproduction on microfilms or in any other physical way, and transmission or information storage and retrieval, electronic adaptation, computer software, or by similar or dissimilar methodology now known or hereafter developed.

Trademarked names, logos, and images may appear in this book. Rather than use a trademark symbol with every occurrence of a trademarked name, logo, or image we use the names, logos, and images only in an editorial fashion and to the benefit of the trademark owner, with no intention of infringement of the trademark.

The use in this publication of trade names, trademarks, service marks, and similar terms, even if they are not identified as such, is not to be taken as an expression of opinion as to whether or not they are subject to proprietary rights.

While the advice and information in this book are believed to be true and accurate at the date of publication, neither the authors nor the editors nor the publisher can accept any legal responsibility for any errors or omissions that may be made. The publisher makes no warranty, express or implied, with respect to the material contained herein.

Managing Director: Welmoed Spahr
Editorial Director: Todd Green
Acquisitions Editor: Jonathan Gennick
Development Editor: Laura Berendson
Technical Reviewer: Richard Conway
Coordinating Editor: Jill Balzano
Copy Editor: Mary Behr
Compositor: SPi Global
Indexer: SPi Global
Artist: SPi Global
Cover image designed by Freepik

Distributed to the book trade worldwide by Springer Science+Business Media New York, 233 Spring Street, 6th Floor, New York, NY 10013. Phone 1-800-SPRINGER, fax (201) 348-4505, e-mail orders-ny@springer-sbm.com, or visit www.springeronline.com. Apress Media, LLC is a California LLC and the sole member (owner) is Springer Science + Business Media Finance Inc (SSBM Finance Inc). SSBM Finance Inc is a **Delaware** corporation.

For information on translations, please e-mail rights@apress.com, or visit http://www.apress.com/rights-permissions.

Apress titles may be purchased in bulk for academic, corporate, or promotional use. eBook versions and licenses are also available for most titles. For more information, reference our Print and eBook Bulk Sales web page at http://www.apress.com/bulk-sales.

Any source code or other supplementary material referenced by the author in this book is available to readers on GitHub via the book's product page, located at www.apress.com/9781484221426. For more detailed information, please visit http://www.apress.com/source-code.

Printed on acid-free paper

To the people who mean the most to me:
my wife, my children, my parents, and my family.

Contents at a Glance

Contents

About the Author

Scott Klein is a Microsoft Senior Program Manager with a passion for Microsoft's data services and technologies. He spent the four previous years traveling the globe evanglizing SQL Server and big data, and he spent the last year working with and sharing his excitement for Microsoft's IoT, Intelligence, and Analytics services, so much so that he can frequently be found buried underneath a pile of Raspberry Pis and other devices. Not forgetting his roots, he still works with SQL Server to the point that he spends most of his day in buildings 16 and 17.

You can find Scott hosting a few Channel 9 shows, including Data Exposed (`https://channel9.msdn.com/shows/data-exposed`), The Internet of Things Show (`https://channel9.msdn.com/Shows/Internet-of-Things-Show`), and SQL Unplugged (`https://channel9.msdn.com/Shows/sql-unplugged`). Scott was one of the four original SQL Azure MVPs, and even though they don't exist any more, he still claims it. Scott thinks the word "grok" is an awesome word, and he is still trying to figure out how to brew the perfect batch of root beer. To see what Scott is up to, follow him on Twitter at @SQLScott or on his blog at `http://aka.ms/SQLScott`.

About the Technical Reviewer

Richard Conway has been programming since his ZX81 days through a morass of jobs in a number of verticals and some spectacularly failing startups. He is a Microsoft Regional Director and Azure Most Valuable Professional with a penchant for all things cloud and data. He is a founder and director of Elastacloud, a cloud data science consultancy based in London and Derby, and he is a founder and organizer of the UK Azure Group and IoT and Data Science Innovators. His latest project is AzureCraft, a twice-yearly conference for children that teaches about the cloud, data, and AI using Minecraft. Follow him on Twitter at @azurecoder.

Acknowledgments

Apress gave me a page to list all of my acknowledgements. Honestly, I don't think a page will be long enough. The list of people who helped me through this book, provided feedback, added insight and direction, and probably let me bug them way too much is very long. But I'll try.

First and foremost are the awesome, and extremely patient, people at Apress. Jonathan Gennick and Jill Balzano are beyond phenomenal. And patient. Did I mention patient?

Next comes all of my co-workers, the fantastic PMs at Microsoft who build these wonderful services: Elio Damaggio, Ryan Crawcour, Matthew Hicks, Saveen Reddy, Matthew Roche, Ashish Thapliyal, Anand Subbaraj, Sharon Lo, Saurin Shah, Michael Rys, John Taubensee, Konstantin Zoryn, Arindam Chatterjee, Rajesh Dadhia, and a host of others.

Next comes the more-than-amazing technical reviewer, Richard Conway. Richard is a big data, IoT, and analytics rock star. I've known Richard a few years and I could not have been more excited to have Richard review this book. I am pretty sure I will forever be in his debt for all the questions I asked him and for the help I received from him.

Lastly, but most importantly, comes my family. Enough cannot be said about the love and support I received from them. So, please excuse me while I go reintroduce myself to them. ☺

Introduction

I'll cut right to the chase here and not be long-winded. The intent of this book is to provide some insight into how Microsoft's Internet of Things and Intelligence and Analytics services can be used together to build an end-to-end solution. This book takes one example and walks that example through the book, implementing service after service, to help stitch together the end-to-end picture. Along the way, the book will stop and look at other interesting, real-world scenarios to help clarify and broaden the picture, to give you further ideas.

This book does not go deeply into any specific service. Entire books could probably be written on each service covered in this book. That is not the intent of this book. My goal is to discuss each service enough to help you decide how you can use the particular service in an IoT solution. I want to get you excited about working with IoT, its capabilities and possibilities. This area is still in its infancy, and there is so much more that can be done to make the quality of our lives better—not to the WALL-E point because heaven forbid we ever get to that point, but to where we are using IoT to save lives, make things safer, and do great things for this planet and humankind.

I tried my best to keep the screenshots and related information up to date. However, if you work with Microsoft Azure, and probably in cloud service for that matter, you know how fast things can change. If the screenshots changed between when I wrote the chapter and it got printed and into your hands, that's the way working with cloud services is. It should be close enough for you to figure out the differences and work with them.

Almost every author will tell you that when they write a book, they look back and wish they could have added "x" or talked more about "y." Such is the case in this book as well. Not to make excuses, but I had to make decisions on many areas just to get the book out. I couldn't keep holding the book up because feature "x" was coming or improvement "y" was about to be released. Thus, I relied on feedback and input to create what is in your hands. I hope you find it good enough.

There are many things I would have liked to discuss in more detail. For example, I would have liked to discuss more about Azure Stream Analytics sliding windows and spent more time in U-SQL (which probably deserves its own book, by the way). However, to get the book out, I hopefully provided enough information to paint the end-to-end picture I was striving for.

If you want to learn more about these amazing services, especially feature "x" and improvement "y," I'll continue to add to this book via my blog, http://aka.ms/SQLScott. If you have ideas or want more information on a specific topic, feel free to ping me via Twitter at @SQLScott or my blog. I am always interested in your questions and feedback because that makes us both better.

Getting Started

CHAPTER 1

■ ■ ■

The World of Big Data and IoT

If you're reading this book, I would venture to guess it is for one of two primary reasons. First, you *want* to learn and get into the big data/Internet-of-things space and technologies. Second, you *need* to learn about dealing with big data. The first is more from the perspective of curiosity and career growth and you are to be applauded for taking the initiative and first step into the fantastic world of dealing with vast amounts of data. If it's the second, I pray for you because this is a "Holy data explosion, Batman! I need to make sense of all this data!" situation and you have some work ahead of you.

Now, I don't want to scare you into thinking that dealing with big data is scary and overwhelming, so don't hyperventilate. It can certainly appear overwhelming, especially if you look at all the technologies that have emerged over the last few years to deal the exponential growth of data and help solve the data insights problems.

But before we begin the journey into this wonderful space, let's go back in time and look at where things have come from, sort of a "history of data." It wasn't too long ago that data was typically stored in simple text format in flat files in either a comma- or tab-delimited format. Reading and writing data took a fair amount of code. If you're over 40, you remember these days not too fondly. Luckily, the relational database came to the rescue, which made dealing with data much easier. Gone were the days of parsing files and reading one line at time; instead we could write SQL queries and have the processing done more efficiently at the server. The relational database veiled the complexity of the storage layer while at the same time provided built-in relational capabilities that flat files did not.

The fact that relational databases still have value stands as a tribute to their effectiveness and capabilities even when storing and working with terabytes of data. But we need to be realistic. As good and effective as a relational database is, we need to have a clear understanding that today's needs can't be solved by a relational database easily. As the size and different types of data being generated and collected continues to rise, it becomes increasingly obvious that we are moving beyond the capabilities of even the best-of-class enterprise relational database systems.

Another thing to consider is how big data changes how we keep (i.e. store) data. On-premises storage is expensive, but with cloud storage being so inexpensive, companies can hang on to their data much longer. As cheap as cloud storage is, companies can keep every raw data point, from clickstream, weblogs, and telemetry data. Many companies today are storing tens of petabytes of data in the cloud.

Which leads us to the need to understand big data and ultimately the Internet of Things (IoT). In this chapter, I will begin by looking at the overall topic of big data, its characteristics, and how we should think about it. I will use that foundation to discuss the hot topic of the Internet of Things. I'll then wrap up the chapter by looking at several scenarios where both big data and IoT are a common place to provide a visual and foundation for the rest of this book.

© Scott Klein 2017
S. Klein, *IoT Solutions in Microsoft's Azure IoT Suite*, DOI 10.1007/978-1-4842-2143-3_1

Big Data

There is some argument on when the term "big data" was first used or coined, but if you think that the topic is relatively new, you are grossly mistaken. Back in 1941, 76 years ago, the term "information explosion" was first used in the Oxford English Dictionary. So even way back then we were trying to understand the phenomenon of the increasing volume of data. Regardless of when or who came up with the term, the important point is that we now live in a time where data is growing at an exponential rate and we need to understand what it is and how to gain insights into its depths, which is what this section will attempt to do. What follows will provide the foundation of what big data is and what it is not, and how to think about it in terms of your business.

What Is Big Data?

If you stop to think about how application data has evolved over time, you can see what is happening out in the enterprise ecosystem, which ultimately makes us stop and think about how the process of storing and processing data and information has evolved. Figure 1-1 illustrates the progress of both the volume and variety of data into what we know today as the big data era. The X axis represents the complexity and variety of data, and the Y axis the volume of data.

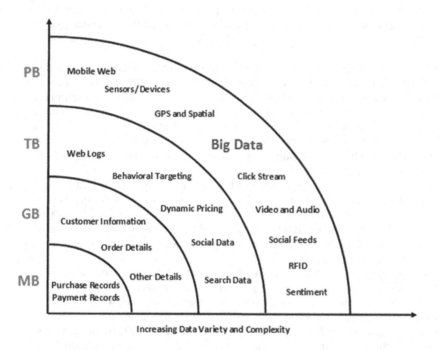

Figure 1-1. *Evolution of application data*

Many of us remember the days when dealing with data meant working in spreadsheets or smaller databases like Microsoft Access or FoxPro. In this era, you dealt with a defined amount and type of information, such as storing types of transactions such as purchases or orders. These scenarios had limited reporting and information insight into the business.

From there, as architectures grew and changed, the need for more advanced data management systems became necessary along with the need to do some level of advanced analytics. For example, a popular scenario was the need to start studying customers and what they were buying. The analytics allowed, for example, businesses to provide purchase recommendations and other opportunities to drive business opportunities.

This scenario, along with more detailed data insight requirements, also brought around more powerful database that added more analytics processing, dealing with cubes, dimensions, and fact tables for doing OLAP analysis. Modern databases provided greater power and additional proficiencies for accessing information through query syntaxes such as SQL. This also abstracted the complexity of the underlying storage mechanism, making it easy to work with relational data and the volume of data it supported.

But there quickly came a point when we needed more than what current enterprise-level relational database systems provided. Exponential data growth was a reality and we needed deeper and distinct analysis on historical, current, and future data. This was critical to business success. The need to look and analyze each individual record was right in front of us, but the infrastructure and current data management systems didn't and couldn't scale. It was difficult to take current systems and infrastructure and apply them to the deeper analytics of analyzing individual transactions.

This isn't to say that the current architecture and systems are going away. There is and will be a need for powerful relational database and business intelligence systems like SQL Server and SQL Server Analysis Services. But as the amount data grew and the need to get deeper insights into that data was needed, it was hard to take the same processing, architecture, and analysis model and scale it to meet the requirements of storing and processing very granular transactions.

The strain on the data architecture along with the different types of data has challenged us to come up with more efficient and scalable forms of storing and processing information. Our data wasn't just relational anymore. It was coming in the forms of social media, sensor and device information, weblogs, and more. About this time we saw the term "No-SQL" bounced around. Many thought that this meant that relational was dead. But in reality the term means "not only SQL," meaning that data comes in all forms and types: relational and non-, or semi-, relational.

Thus, in essence, we can think about big data as representing the vast amounts and different types of data that organizations continue to work with, and try to gain insights into, on a daily basis. One attribute of big data is that it can be generated at a very rapid rate, sometimes called a "fire hose" rate. The data is valuable but previously not practical to store or analyze due to the cost of appropriate methods.

Additionally, when we talk about solutions to big data, it's just not about the data itself; it also includes the systems and processes to uncover and gain insights hidden inside all that wonderful, sweet data. Today's big data solutions are comprised of a set of technologies and processes that allow you to efficiently store and analyze the data to gain the desired data insights.

The Three Vs of Big Data

Companies begin to look at big data solutions when their current traditional database systems reach the limit of performance, scale, and cost. These big data solutions, in addition to helping overcome database system limitations, also provide a way to more effectively provide the much-needed avenue for gaining the greatly required insight into the data to further examine and mine data in a way that isn't possible in database systems. As such, big data is typically described and defined in terms of the three Vs to understand big data and solve big data issues: volume, variety, and velocity.

Volume

The volume of data has reference to the scale of data. Where relational database systems work in the high terabyte range, big data solutions store and process hundreds or thousands of terabytes, petabytes, and even exabytes, with the total volume growing exponentially. A big data solution must have the storage to handle and manage this volume of data and be designed to scale and work efficiently across multiple machines in a distributed environment. Organizations today need to handle mass quantities of data each day.

Variety

Data today comes in many different forms, both structured and semi-structured, relational and non-relational. Gone are the days where the majority of data resides in a relational database. Today, a very large amount of data being generated comes from sensors, devices, social media, and other formats that aren't conducive to a structured or relational format. The key to keep in mind here is that it is just not all about relational data anymore. Today, organizations are dealing with relational (structured) and non-relational (semi-structured or non-structured) data. Typically, the majority of data currently stored in big data solutions is unstructured or semi-structured.

A key factor when dealing with a variety of data is deciding how and where to store the variety of data. Using a traditional relational database system can be challenging and may no longer be a practical solution for storing semi-structured or non-structured due to a lack of a schema application. Big data solutions typically target scenarios where there is a huge volume of unstructured or semi-structured data that must be stored and queried to extract business intelligence.

Velocity

The velocity of data has dual meanings. The obvious meaning applies to how fast data is being generated and gathered. Today, data is being produced and collected at an ever-increasing rate from a wide range of sources, including devices and sensors as well as applications such as social media.

The second meaning of velocity applies to the analysis of the data being streamed in. Businesses and organizations must decide how fast they need to understand the data as it is being streamed in.

To help put these Vs in perspective, here are some examples:

- Microsoft Bing ingests over 7 petabytes of data a month.

- The Twitter community generates over 1 terabyte of tweet data every single day.

- Five years, ago, it was predicted that 7.9 zettabytes of data would make up the digital universe.

- 72 hours of video are uploaded per minute on YouTube. That's 1 terabyte every 4 minutes.

- 500 terabytes of new data are ingested in Facebook databases.

- Sensors from a Boeing jet engine create 20 terabytes of data every hour.

- The proposed Square Kilometer Array telescope will generate "a few exabytes of data per day" per single beam.

I'll give you another example that will help put the data explosion into context. In 1969, the United States wanted to put a man on the moon. Keep in mind this was 48 years ago and it had to work the first time, meaning that we wanted to send a man into space and we also wanted to get him back. So, NASA built Apollo X1 with a weight of almost 30,000 pounds (13,500 kg) and a top speed of almost 2,200 mile per hour (3,500 km/hour). The distance to the moon is about 221,208 miles (356,000 km). NASA got the brightest minds together and when it was all said and done, it took 64Kb of RAM with the code written in Fortran to get Neil Alden Armstrong to the moon and back.

Today, 47 years later, the health messages alone from the game Halo generate gigabytes of data per **second**. PER SECOND! A GAME! Yet it only took 64Kb of RAM to send man to the moon. Mind blowing.

Are There Additional Vs?

While it is universally agreed upon that volume, variety, and velocity make up the three main Vs of big data, some have added additional Vs to the big data problem. There is some disagreement, however, as to what they are. Some list value as the fourth V, while others list veracity as the fourth V. Some still list all way out to seven Vs. Personally, I think that if you're all the way out to seven Vs, you're just looking for words that start with the letter V. Yet, I do find worth in *veracity* and *value* so I will briefly discuss them here.

- **Veracity**: This term has reference to the *uncertainty* of your data. Essentially, how accurate is your data? It's worthless if it's not accurate. You can't make trusted and accurate decisions if you don't trust your data, and your programs are only as good as the data they are working with.

- **Value**: Big data is data that has value, and having access to this data does no good unless you can turn it into something of meaning.

Other Vs have been mentioned, such as variability. However, what's important to you and me is to understand how these Vs help provide insight into the scale and importance of data, including the challenges of dealing with big data.

Why You Should Care About Big Data

Organizations today that deal with the Vs discussed above look to big data solutions to resolve the limitations of traditional database systems. Today's requirements for storing data are outpacing those of a relational data store, and businesses today may not survive into tomorrow without the enabling power and flexibility of big data solutions.

The data warehouses of yesterday are being outgunned by the big data solutions of today and tomorrow simply because the volume of data surpasses the cost and capacity found in relational database systems. That doesn't mean that data warehouses are passé and not needed, but in fact quite the opposite. Organizations start caring about big data solutions as a new way to gain rapid and valuable insights into their growing data when the volume, velocity, and variety of data exceeds the cost and capabilities of a data warehouse storage solution. One does not replace the other; they are complimentary.

The time to start caring about big data is when the Vs of big data are a reality within your organization: when your data is no longer just relational; when you have the need to store and process up to petabytes of data in a cost-effective way; when you have the need to find valuable insights into that ever-increasing volume of data quickly and efficiently.

Big Data Solutions vs. Traditional Databases

As the landscape grows and progresses from traditional database systems to big data solutions, it is helpful to understand how traditional database systems and solutions differ from big data solutions. Table 1-1 summarizes the major differentiators and provides a high-level comparison.

Table 1-1. *Comparing Relational Database Systems to Big Data Solutions*

	RDBMS	Big Data Solutions
Data Size	Gigabytes (terabytes)	Petabytes (hexabytes)
Data Types	Structured	Semi-structured or unstructured
Access	Interactive and batch	Batch
Update pattern	Read/write many times	Write once, read many times
Structure	Static schema	Dynamic schema
Integrity	High (ACID)	Low
Scaling	Nonlinear	Linear

Traditional database systems use a relational design where data is stored and based on predetermined schema, and are designed to handle workloads up to terabytes in size. These systems are meant for OLTP (Online Transaction Processing) transactions, meaning that applications are reading, writing, and updating data within the database at a high frequency, usually using small transactions that affect a few rows at a time.

Scaling a relational database to gain performance needs typically means a nonlinear scalability model, meaning that you are either adding more memory or CPU (or both) to the node where all the transactions take place, and/or applying disk partitioning and filegroups that divide the data into multiple logical chunks, with those chunks residing on the same physical node or storage system, or different physical nodes and storage systems resulting in network latency.

In contrast, big data solutions allow you to store any type and format of data including non-structured and semi-structured data while not requiring nor operating over a predetermined schema. Together, these two features allow data to be stored in its native, raw format and apply a schema only when data is read.

While both traditional databases and big data solutions store and query data, one key difference with big data solutions is how data distribution and query processing takes place, and how data moves across the network. Big data solutions work and function in a cluster in which each node in the cluster contains storage and the initial data processing takes place at each node. With the data already loaded onto each node in the cluster, no data needs to be moved onto each node for processing, thus improving processing performance.

Query processing in big data solutions are primarily batch operations. With massive data volumes, a variety of data formats and types, and a tendency for queries to be a bit more complex, these batch operations are likely to take some time to yield results. These batch queries, however, run as multiple tasks across the cluster, making it much easier to handle the volumes of data, and as such, provide a level of performance that is typically not seen in traditional database or other systems. And since we're talking about querying, it should be pointed out that in big data solutions, batch-type queries are by-and-large not frequently executed. Compared to traditional databases where queries are regularly executed as part of a process or application, big data queries are much less frequently executed, which becomes much less of a shortcoming.

Hopefully the insights into the differences between traditional databases and big data solutions provided above has helped make clear that these two solutions are complimentary. Relational databases will continue to be around for quite some time. In fact, it is quite common to see the results of a big data query to be stored in a relational database or data warehouse to be used for BI (Business Intelligence) reporting or further down-the-stream processes.

Putting all of this into perspective, it is all about how organizations tackle the problem of the three Vs. Organizations look to big data solutions to solve the classic problem of how to discover the hidden gems of useful and meaningful information in their data.

A Quick Data Landscape Comparison

It wasn't too long ago that the data landscape consisted of only a handful of technologies to understand data. Data mining, data warehouses, and BI have been a staple and go-to for data processing and data insights for a long time. Today, it's impossible to have the same conversation without including data analysis (or analytics) and big data in the same breath. You can also throw IoT into the conversation, but that will be discussed in the next section so it will be left out of this comparison.

What I want to do is provide a quick look at how big data and analytics compare to the technologies that many data professionals have been using to those which are more recent, specifically BI, data mining, analytics, and big data. Let's not get hung up on the term "data professional" either. For the sake of argument, a data professional can mean a SQL DBA, SQL developer, data scientist, or BI architect.

As a note, this is not a deep, analytical comparison, but more of a simple discussion toward understanding the boundaries between each area and technology, as these tend to blur when they come up in discussion. In fact, you will probably have different opinions, and that is fine. The goal here is simply to help understand the different concepts and navigate their relationships between each other.

- **Business Intelligence**: BI is data-driven decision making based on the generation, aggregation, analysis, and visualization of data. Discussions about BI go beyond the data into what insights can be gleaned from it. BI is spoken in terms of both technology, such as data transformation (ETL processes), and reporting, as well as general processes that encompass the technologies that support the processes.

- **Data Mining**: Data mining is the process of discovering actionable information from large sets of data. It is sifting through all the evidence looking for previously recognized patterns, and finding answers you didn't know you were looking for.

- **Analytics**: Analytics focuses on all the ways you can break down the data and compare one nugget of data to another, looking for patterns, trends, and relationships. Analytics is tied at the hip with BI. Analytics is about asking questions; BI is about making decisions on those questions. Machine learning is a type of analytics—predictive analytics, asking questions about the future.

- **Big Data**: Solutions and technologies that store and process massive volumes and different types of structured and semi- or unstructured data.

There are certainly similarities between each of these terms; for example, one could argue that analytics and BI are synonymous, or rely on one to the exclusion of the other. What helps blur the line is the great tooling available, such as Power BI. It is a tool that provides both data analysis and data visualization.

Internet of Things (IoT)

A portion of the first section talked about the amount and different types of data being generated and the ways to handle that data, and I also listed some examples of the different types of applications and devices that are generating that data (telescopes, jet engines). In order to understand big data, it helps to understand what is generating the different types and amount of data, and that is what this section is about: the "things" that are generating the data and passing that data around the Internet. Thus, the Internet of Things.

The term "Internet of Things" was coined by the guy who helped create RFID, Kevin Ashton. In fact, it was lipstick that ultimately led to the coining of the IoT term. At the time, Kevin was employed at Procter & Gamble as the Oil of Olay lipstick brand manager, and he noticed that a popular color of their lipstick was continually out of stock. Kevin decided to find out why and in digging into the problem he discovered that there was a data problem between the Procter & Gamble stores and the supply chain. The solution he came up with led him to drive the development and deployment of RFID chips on inventory. Kevin asked himself "What if I took the radio microchip out of the credit card and stuck it in my lipstick? Could I then know what was on the shelf, if I had this shelf talk to the lipstick?"

Kevin's efforts in RFID development led to him being loaned out to MIT to start the technology group Auto-ID Center, which would continue the research of RFID technology. It was during this time that he coined the phrase "Internet of Things" in 1999. As such, Kevin is known as the "Father of the Internet of Things."

What Is the IoT?

Simply put, the Internet of Things refers to devices that collect and transmit data over the Internet. These devices can be anything from your toaster or washing machine to your cell phone or wearable devices such as the Microsoft Band, Apple Watch, or Fitbit, just to name a few. It is said that if a device has an on/off switch, generates data, and can connect to the Internet, chances are that it can be part of the IoT. Today, many cars are connected to the Internet. By 2020, it is estimated that over 250,000 cars will be connected to the Internet. The wearable device market grew 223% in 2015 globally. Many companies are investing in home devices, such as Samsung. It is estimated that by "connecting" kitchens to the Internet, the food and beverage industry could save as much as 15% annually. According to some estimates, the Internet of Things will add $10-$15 trillion to the global GDP in the next 20 years.

Other popular devices that are used today to generate data are the low cost, credit card-sized computers such as the Raspberry Pi, Pi Zero, and Tessel IO boards. These little devices are frequently used to create and generate data in places and situations where normal PCs and laptops can't go or aren't efficient enough. In fact, it is these devices that will be used in this book to generate data.

However, even though the term Internet of Things was coined in 1999, the concept of collecting and transmitting data goes back earlier than that. ATMs did it as far back as 1974. Still, with the recent movement and excitement around big data and IoT, IoT is still a new concept to a lot of people.

There are several key factors that are helping drive the IoT space, and one is the inexpensive costs of the components. For example, both the Raspberry Pi and Tessel IO board are less than $40, and the associated sensors and modules, such as GPS, accelerometer, climate, and ambient modules, are even less. Each of these credit card-sized boards is powerful enough to run advanced software, such as Windows 10 Core. There are other boards as well, including ones from Intel and Arduino. Chapter 2 will cover these boards.

Another factor is the advancement in software that provides rich, dynamic, and high-level data processing and analysis capabilities. The software is becoming more powerful and easy to use, providing the ability to deliver the high-level and enterprise-class analytics big data solutions need.

The explosion of cellular and wireless connectivity has also boosted IoT solutions, allowing big data and IoT solutions to include mobile components and provide connections to the Internet where previously impossible. As connectivity to the Internet continues to improve and become less of a cost barrier, the increase in IoT scenarios has accelerated tremendously.

Lastly, a major key factor in the explosion of IoT scenarios has been the rapid advancement in cloud services and technologies, providing a highly cost-effective solution for data storage, processing, and analysis. Coupled with the fact that the current development tools and technologies used on-premises also work when developing for the cloud makes using cloud-based services and solutions highly advantageous.

The Internet of "Your" Things

It was mentioned earlier that the Internet of Things refers to devices that collect and transmit data over the Internet. This year it is estimated that there will be over 5 billion connected "things," or devices, to the Internet. Gartner estimates that by 2020 there will be over 25 billion connected devices, and some believe that the number will be even higher. Today, businesses are using the data generated by these devices to create tremendous business value through the analysis of this data.

The question you need to ask yourself is, what does IoT mean for you? As much talk as there is about IoT, it is useless unless the discussion includes the things, devices, and data that are important for *your* business. Thus, the discussion is really about the Internet of *Your* Things: the devices and data that have an influential impact on unlocking the value and insights your business needs.

The sole purpose of this book is to clearly lay out Microsoft's view on IoT and help paint a clear picture of Microsoft's IoT ecosystem of services and technologies. Thus, the chapters in this book will walk through the generation of data via devices and sensors, and the consumption and analysis of that data using Microsoft's suite of cloud services and technologies to gain insight and analysis of the generated data.

The first set of chapters will focus on "data on the move," meaning data that is being generated, processed, and analyzed in real time. The second set of chapters will focus on the processing and analysis of data at rest, meaning data that has been consumed from devices and sensors and stored for future analysis. Together, the intent of these chapters is to provide an end-to-end example of how one might use Microsoft's services to gain real-time insight into the data generated by the Internet of "Their" Things.

Before proceeding, however, the next few pages will take a look at several real-world examples used today. You may have heard about them before but they are mentioned here because they provide insight and examples into IoT solutions that are in place today.

Scenarios

There are a plethora of scenarios in which IoT solutions have been created and used to improve and solve countless data problems. This section will discuss a few of the scenarios to help paint a clear picture as to what is possible with IoT solutions as well as provide a solid foundation of what the chapters in this book will provide.

The Connected Car

There is commercial (or advert) in the United States by a popular insurance company that has been on the television for a while. The premise of the commercial simply states that you can save on your auto insurance through safe driving habits. The way it works is this: if you are signed up with their insurance and you sign up for this program, they will send you a little device that you plug into your vehicle's OBDII diagnostic port. When the vehicle is on, this device sends information about your trip to the insurance company. This information includes information such as how fast you accelerate, how hard you brake, speed, time of day you drive, etc. All of this information is sent to the insurance company to determine if your driving habits could turn into a cheaper rate. There may be examples of this in other countries as well with other insurance companies.

Personally, this is too "big brother" for me. Are they going to raise rates if they don't like the way I drive? The next step might be putting a speaker in the device and using cloud-to-device messaging (which I'll discuss in the next chapter or two) to tell me in real time that it noticed I was a bit heavy on the gas and to slow down. Nope, I don't need that (not that I am a speeder ☺).

All kidding aside, this is a very simple but excellent example of a connected car. Data is being generated by a device and sent to a location for real-time or near-real-time analysis. In the bigger picture, think of the number of these devices out on the road and the data being generated by them and sent to the insurance company. Now think about how the data is being used and analyzed.

This data could be very beneficial to not only the insurance company but to auto manufactures and other companies. Taken together, all of this data could be used to make cars safer or prevent accidents from happening by preemptively diagnosing an engine problem. From a predictive analysis perspective, possible accidents could be avoided by looking at people's driving patterns and habits.

Another example is Formula 1. Using the cloud, many race teams use real-time data analysis during race time to finish. Formula 1 cars have hundreds, if not thousands, of sensors on their cars, and today these sensors send data back to the pits in real time. This data is then displayed on a dashboard back in the pits for real-time analysis, making for more efficient pit stops.

More and more car manufacturers are building cars that enable people to perform remote diagnostics on their cars and send the data to the Internet for insight and analysis. Many automakers today have made 4G wireless connectivity available in new cars, paving the way for a plethora of services and information, including better navigation, real-time traffic and parking information, as well as enhanced rider experiences.

Mercedes-Benz has recently introduced models that can link to Nest (`https://nest.com/`), the IoT-powered smart home system which can remotely activate a home's temperature. Cars in the United States and Canada have an ODB (onboard diagnostic) port, which has been mandatory since 1996. It is through these ODB ports that a rapidly growing business of cloud-connected mobile apps are springing up, providing myriad services including maintenance reminders, diagnostic access, and much more.

Connected Home

The advancement of connected home technology has gained a lot of momentum lately. The company X10 (`www.x10.com/`) has provided home automation gadgets for almost 40 years now. What started as just gadgets and widgets has morphed into modern versions now offered by X10, for example, as well as fun, simple, yet critical projects such as a simple home security system that uses a Kinect and Microsoft's cloud services, made by good friend and co-worker Brady Gaster (go to `https://github.com/bradygaster/ Kinectonitor`).

Other examples of connected homes are things like the June Intelligent Oven or the Samsung Family Hub refrigerator, both of which let you see what's going on inside from your smartphone.

The advancement of IoT solutions in the area of connected homes and buildings continues to focus on controlling nearly every aspect including remote diagnostics, maintenance, and analytics. The connected home is such a growing industry that it warrants its own conference, the Connected Home and Building conference (`www.connectedhomecon.com/east/`).

I recently read in a March 2016 Business Insider article that connected home device sales will drive over $61 billion in revenue for 2016, with 52% compound annual growth rate to reach $490 billion by 2019. That same report stated that the connected home device category makes up roughly 25% of shipments in the IoT category, and connected home device shipments will outpace smartphone or tablet shipments. That's significant.

Let's not, however, confuse *smart home* with *connected home*. A smart home provides the ability for me to turn the lights on from my phone. A connected home is about the data the devices in your home are generating and how that data is being used to improve home life, save money, drive new products, and keep people safe.

Connected Cow

The connected cow example is a fun and interesting scenario. Perhaps you have heard it, but if you haven't, have a seat. I have talked about connecting cars and homes, but farm animals? This is a scenario where technology has transformed even the oldest industry.

The connected cow scenario is simply where pedometers were strapped to the legs of cows. Each pedometer was connected to the Internet and simply sent the step count of each cow to a web dashboard back in the farmhouse via Microsoft Azure.

Why was this being done? A farm in Japan was looking to solve two things. First, the farm wanted to detect health issues early and prevent herd loss. Second, the farm wanted to improve cattle production by accurately detecting estrus. Estrus is when the animal, in this case the cow, goes into heat.

The problem was that cattle go into heat for a very small window of time (12-18 hours) every 21 days, and this occurs mostly between 10pm and 8am. The goal was to use technology to more accurately detect the estrus period during which time artificial insemination would take place, so that the pregnancy rate would significantly increase.

Without technology, managing hundreds or thousands of cows and accurately detecting when each cow goes into heat is nearly impossible. With technology, this could be doable, and in actuality, it was.

The farm in Japan contacted Fujitsu for help. Fujitsu created a solution that used pedometers. What they found was that a cow takes many more steps when it goes into heat. So they strapped a pedometer to every cow. These pedometers sent the step count to a Microsoft Azure solution, which analyzed the data and sent alerts.

Using this solution, they were able to get up to 95% accuracy in the detection of estrus as well as the optimum time for artificial insemination. Upon further analysis, Fujitsu and the researchers also found out something else very interesting. They found out that there is a window around the optimal timeframe for artificial insemination. They found that if you perform artificial insemination in the first half of the window, you are more probable to get a female (cow), and if you perform artificial insemination in the second half of the window, you are more probable to get a male (bull), with up to 70% probability. The farmer now has the ability to control production whether he needs more cows or bulls.

Another interesting thing they found is that they can detect from between 8 and 10 different diseases from the step patterns and number of steps returned from the pedometers of the cows.

Also, the farm was able to improve their labor savings simply from not needing to manually monitor the cows constantly. The impact of savings from implementing the solution was quite significant in many aspects.

There are many great examples of how IoT is being used today, and we are just scratching the surface in each of these areas, so there is more to explore and discover. As you explore and learn more about IoT, think about how all of this data can be used to improve our way of life, lower costs, save lives, and much, much more.

Summary

Today it seems like everyone has their own definition of what big data and the Internet of Things means. But there is no question that there is huge opportunity and a lot of potential in this space. This chapter began by defining big data and looking at the characteristics of big data. To help lay the foundation for the rest of the chapter, and the book, the chapter discussed why understanding big data concepts and technologies is important.

Building on the big data theme, the chapter shifted from big data to the Internet of Things, first providing insight into where the term comes from and then taking a good look into what IoT is and means. To provide some context, the chapter finished with several scenarios where the Internet of Things is being used to make the world a better place and set forth a challenge to think about how one might use IoT to change the world for the better.

CHAPTER 2

■ ■ ■

Generating Data with Devices

The Internet of Things, or IoT, is about devices that generate and transmit data over the Internet. As mentioned in the last chapter, it is estimated that by 2020 there will be over 25 billion "things" (i.e. devices) connected to the Internet; these devices will range from the washer or oven in your house to the watch you wear or the phone in your pocket. The last chapter also covered several scenarios in which the IoT and devices were implemented, including cars, homes, and animals.

This chapter will simply build on that information and show several examples of the types of devices used today, how to use them to generate data, and how to send that data over the Internet and store it for analysis. The devices that this chapter will use will be the Raspberry Pi board from the Raspberry Pi Foundation (www.raspberrypi.org/) and the Tessel board from the Tessel Project (https://tessel.io/). There are a large number of devices available that provide very similar functionality, including the Adafruit Feather HUZZAH, the Edison from Intel, the DragonBoard from Arrow, and the boards from Beaglebone, but it would be unrealistic to discuss and demo them all here. Visit https://azure.microsoft.com/en-us/develop/iot/get-started/ for a full list of IoT boards supported by Microsoft.

Adafruit makes a great Azure IoT starter kit complete with the Feather HUZZAH board, a DHT22 sensor, cables, breadboard, a power cable, and other jumpers and switches. It is available via the Adafruit web site: www.adafruit.com/products/3032. This chapter won't cover how to use it but I have blogged about how to get it set up and running with Azure. Visit my blog for more information.

The reason I have chosen the Raspberry Pi is because of its ability to run Windows 10 IoT Core and the familiar development environment of Visual Studio, which makes it easy to code and deploy solutions. I have chosen to also discuss the Tessel board not only because it is red (my favorite color) and looks cool, but to also illustrate Microsoft's support for open source technologies. In all fairness, I am including the Tessel in this chapter simply to show Microsoft's support for a wide range of operating systems, languages, tools, and frameworks.

Now, a couple of things by way of disclaimer. First, as discussed in the introduction, this book is about Microsoft's vision of big data and IoT via the IoT and Cortana Intelligence suite of Azure services, thus data generated will be routed as such. However, for the sake of the examples of this chapter, the data generated from these devices will simply be routed to the screen for output. Chapter 3 will show in detail how to create, configure, and connect the device to an Azure IoT Hub for data storage and processing, and Chapter 4 will then hook up the devices to an Azure IoT Hub. This chapter is all about using devices to generate data.

So, with that, let's get started.

Raspberry Pi

A popular IoT device, the Raspberry Pi is a small but proficient device that does everything a normal computer does by plugging in a monitor, keyboard, mouse, and Ethernet cable. You can browse the Internet, play games, or even run desktop applications such as Microsoft Word or Excel. However, this tiny but powerful mini-pc is really targeted to those who want to explore the world of devices and maker projects. A "maker" is a hardware hacker, someone who likes to build things that make the world a better place.

© Scott Klein 2017
S. Klein, *IoT Solutions in Microsoft's Azure IoT Suite*, DOI 10.1007/978-1-4842-2143-3_2

In Figure 2-1 you can see two Raspberry Pis. The device on the top is the Raspberry Pi 2 Model B, which has been available since February of 2015. The device on the bottom is the Raspberry Pi 3 Model B, which became available in February of 2016.

Figure 2-1. *Raspberry Pi 2 (top) and 3 (bottom)*

The Raspberry Pi is about the exact same size as a credit card and both the Pi 2 and Pi 3 vary very little in functionality. Table 2-1 details the main differences between the Pi 2 and the Pi 3.

Table 2-1. *Comparing the Raspberry Pi 2 and 3*

Raspberry Pi 2	Raspberry Pi 3
• 900MHz quad-core ARM Cortex – A7 CPU	• 1.2GHz 64-bit quad-core ARMv8 CPU
• 1GB RAM	• 1GB RAM
	• 802.11n Wireless LAN
	• Bluetooth 4.1
	• Bluetooth Low Energy (BLE)

Both the Pi 2 and the Pi 3 include the following:

- 4 USB ports

- 40 GPIO pins

- Full HDMI port

- Ethernet port

- Camera interface (CSI)

- Display Interface (DSI)

- Micro SD card slot

- Combined 3.5mm audio jack and composite video

Honestly, having the latest and greatest is cool, but if you don't plan on doing any Bluetooth or Wi-Fi, the Pi 2 is perfect for getting started. However, you can't argue $35, so feel free to spring for the Pi 3 if you are so inclined. If you already have a Pi 2 and are looking at the Pi 3, the Pi 3 is identical in form factor so any case you have will work for the Pi 3. Plus, the Pi 3 is completely compatible with the Pi 2.

With the Raspberry Pi in hand, it is time to start setting up and coding. The next section will walk through the setup and configuration of the Raspberry Pi.

Getting Started

Hopefully you noticed on the list of items *not* included was a hard drive. If there is no hard drive, how does it boot up and work? Great question! In order to get the Pi to work, you need a MicroSD card, shown in Figure 2-2.

Figure 2-2. *MicrosSD card*

You can pick up a MicroSD card anywhere. The one in Figure 2-2 is 64GB but you really don't need one that big. An 8GB or 16GB one will do just fine, and you can find them online quite easily. I ordered mine from Amazon:

www.amazon.com/SanDisk-Ultra-Micro-SDHC-16GB/dp/9966573445?ie=UTF8&psc=1&redirect=true&ref_=oh_aui_detailpage_o01_s00

Simply slide the card into the MicroSD card slot on the back of the Pi, as shown in Figure 2-3.

Figure 2-3. *Inserting the MicroSD card into the Raspberry Pi*

However, there is nothing on the MicroSD card, thus to get started with the Pi and for it to be useful, you need to install an operating system on it. For the Raspberry Pi, you'll be putting Windows 10 IoT core on it. You'll need a computer with an SD or MicroSD card slot, so slide your new card into the slot.

The Pis in the figures are shown in a case. Depending on where you order your Pi from, it may not come with a case so please take precautions with anti-static bags or gloves to avoid blowing the electronics on the board if you do not have a case. I ordered these clear cases from Amazon:

www.amazon.com/gp/product/B00MQLB1N6/ref=oh_aui_detailpage_o05_s00?ie=UTF8&psc=1

It is best practice to format your SD card before installing any OS on it, and there is an easy tool which does that. The following website has a nifty utility which formats all sorts of memory cards:

www.sdcard.org/downloads/formatter_4/

Scroll down towards the bottom of the page and click the blue "Download SD Formatter for Windows" link. On the EULA page, scroll down and click Accept and then the download will begin. Unzip the Setup file and install the formatter. Once installed, run the formatter. It will find your SD card. By default the Quick Format is selected, which is fine. Click Format. Within a matter of seconds your SD card will be formatted.

Installing Windows IoT Core

There are several ways to do this, depending on if you have a Pi 2 or Pi 3. I'll begin with the Pi 2.

Pi 2

The easiest way to install Windows 10 IoT Core on your Raspberry Pi 2 is by installing the Microsoft IoT Core Dashboard, which you can download from here:

```
http://ms-iot.github.io/content/en-US/Downloads.htm
```

Once on the Downloads and Tools page, simply click the big blue "Get IoT Core Dashboard" button. It is a small download, so it doesn't take that long. Once downloaded, run the setup program. The install itself is also quick, and once done the IoT Dashboard will automatically launch.

On the left of the IoT Dashboard the "Set up a new device" option should automatically be selected, so click the blue "Set up a new device" button on the main page. See Figure 2-4.

Figure 2-4. *Microsoft IoT Core Dashboard*

On the "Set up a new device" page, make sure the device type is set to Raspberry Pi 2 and that Windows 10 IoT Core for Raspberry Pi is selected. At the time of this writing, Windows 10 IoT Core for Raspberry Pi 2 is the only option but this may change to include Windows 10 IoT Core for Raspberry Pi 3 in the future.

However, clicking the drop-down for device type shows four options: Raspberry Pi, Minnowboard Max, Qualcomm Dragonboard, and Custom, which shows the breadth of devices that Microsoft and Windows 10 IoT Core supports.

Check the "I accept the software license terms" button and click the "Download and install" button, as shown in Figure 2-5.

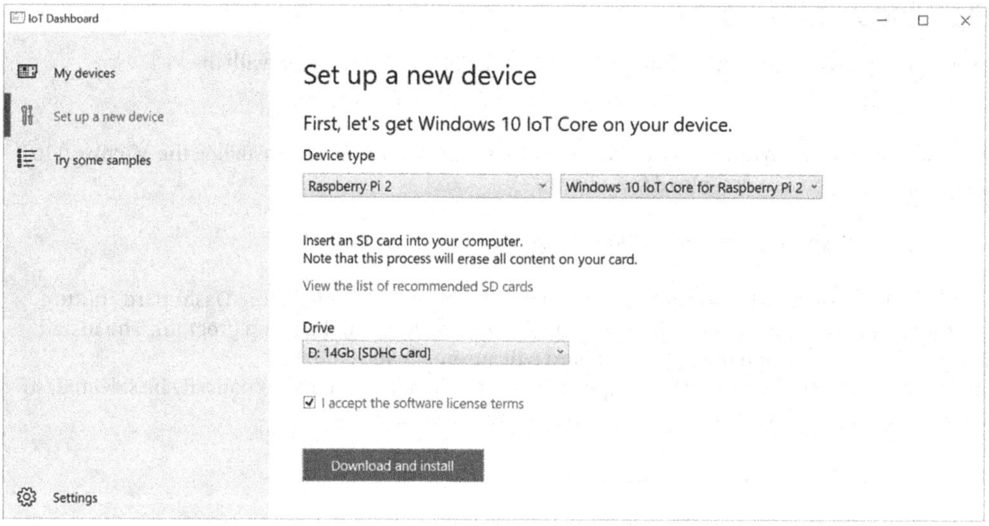

Figure 2-5. *Setting up Windows 10 IoT Core on a Raspberry Pi 2*

You might get a pop-up encouraging you to back up any files on the card before proceeding. Click OK. The Dashboard will flash your device, and then download and install Windows 10 IoT Core to the SD card. See Figure 2-6.

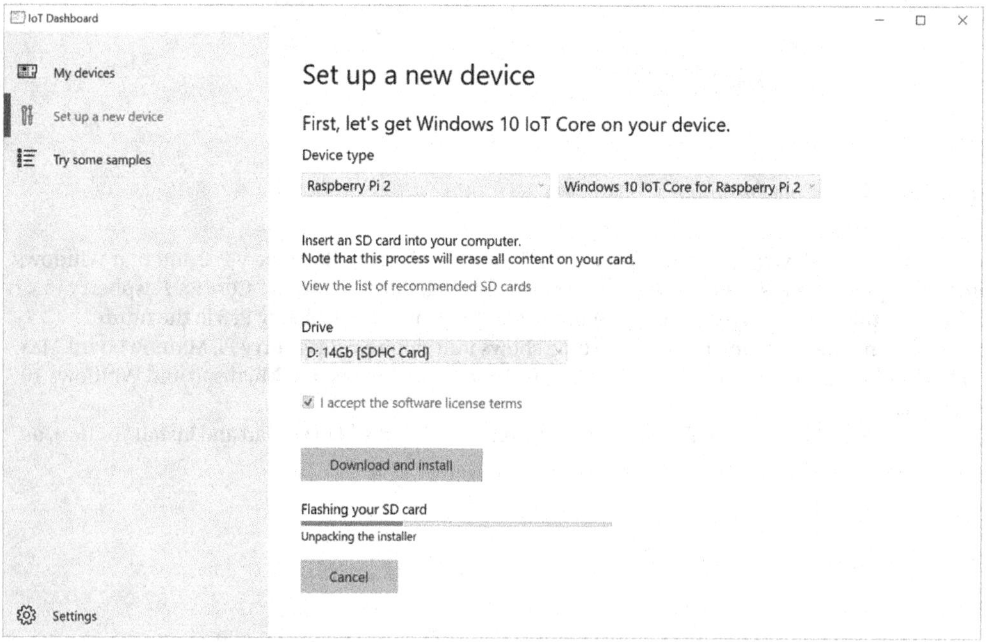

Figure 2-6. *Installing Windows IoT Core for Raspberry Pi 2*

Depending on your Internet connection this might take a few minutes, so be patient (see Figure 2-7).

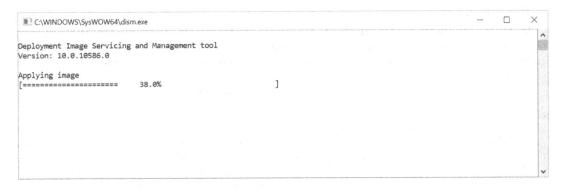

Figure 2-7. *Installing Windows 10 IoT Core on the MicroSD Card*

When the process is completed, your SD card will be ready, as shown in Figure 2-8. Remove it from your computer and insert it into the MicroSD slot in the Raspberry Pi, as shown in Figure 2-3 above. If you want to set up another device, click the "Set up another device" button; otherwise, close the IoT Dashboard.

Your SD card is ready.

1. Insert your SD card into the device

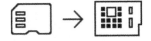

2. Get Connected

□ **Ethernet** (recommended)
Connect your Ethernet cable to your local network and boot up your device

⍡ **Wi-Fi**
Plug in your Wi-Fi adapter and boot up your device.
See a list of supported Wi-Fi adapters

3. Find your device

Note: It will take a few minutes for your device to boot and appear in "My Devices"

Set up another device

Figure 2-8. *Completed process*

At this point you are ready to connect and boot up your Raspberry Pi.

See the next section for the Pi 3 setup.

Pi 3

By the time you read this, you will probably be able to install Windows 10 IoT Core via the Dashboard. However, this section will walk you through another option because currently it is the only way to install Windows 10 IoT Core on a Pi 3. Plus, this other option provides a good look at the Raspberry Pi NOOBS (New Out-of-the-Box Software).

The first thing to do is to go to the Raspberry Pi website and download NOOBS:

`www.raspberrypi.org/help/noobs-setup/`

Click the Download Zip link for the Offline and Network Install. The current version of NOOBS as of this writing is 1.9. Extract the contents of the zip file to a folder, and then drag and drop all the contents onto your MicroSD card. When the copy is finished, safely remove the SD card and insert it into your Raspberry Pi.

Plug in your keyboard, mouse, video, and Ethernet, and then power up your Raspberry Pi by either plugging it in to your PC (via the USB-to-micro-USB cable) or via the power adapter. When the Pi boots up, you will be presented with the screen shown in Figure 2-9. This screen shows you all of the operating systems supported by the Raspberry Pi and which you can install on the SD card. Select the Windows 10 IoT Core option.

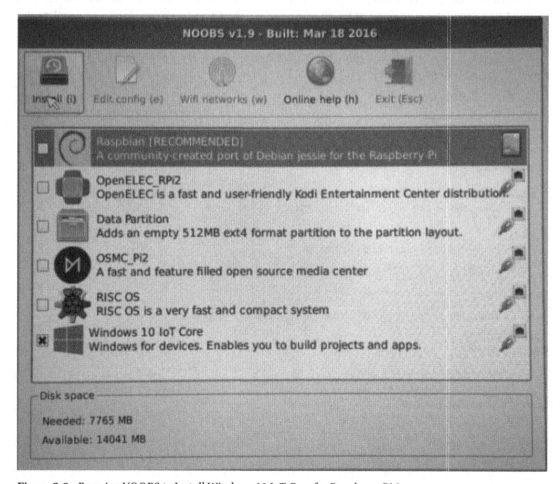

Figure 2-9. *Running NOOBS to Install Windows 10 IoT Core for Raspberry Pi 3*

Click Install. You will be prompted with a warning which states that the selected operating system will be installed and all existing data on the SD card will be overwritten, at which point NOOBS will begin by preparing the operating system partition for use on the SD card and the process for Windows 10 IoT Core will be started.

If this is the first time you are going through this process, you will be asked to provide your Microsoft account login credentials and join the Microsoft Insider Program. Once logged in, you will see a screen asking you to select the Windows 10 IoT core edition. You might see two editions listed:

- Windows 10 IoT Core

- Windows 10 IoT Core Insider Preview

If installing on Pi 2, select Windows 10 IoT Core. If installing on Pi 3, select Windows 10 IoT Core Insider Preview.

Confirm the device you are installing the OS on (Raspberry Pi) and click Confirm, and then click the Download Now button. You will need to accept the license agreement, so click Yes. The OS will then be downloaded and installed on your SD card. Depending on your Internet connection, this might take several minutes. Once downloaded, you will see a confirmation dialog. Click OK to reboot your device. Once the device reboots, you will see the Windows boot screen followed by a Device Info page.

Now you are ready to start putting the device to work and generate some data.

Generating Data

All the devices, including the Raspberry Pi, can be used in many different environments and scenarios. This example will use a simple temperature and humidity sensor to generate data. For this example, I am using both a DHT11 sensor and a DHT22 sensor. You will also need a breadboard and jumper wires. I ordered a sensor kit from SunFounder that contains a total of 37 modules, including the DHT11 sensor, plus the breadboard and jumper wires. I got mine from here:

www.amazon.com/SunFounder-Sensor-Kit-Raspberry-Extension/dp/B00P66XRNK?ie=UTF8&psc=1&redirect=true&ref_=oh_aui_detailpage_o04_s00

You can also order a standalone DHT22 temperature and humidity sensor, and I ordered mine from PrivateEyePi at http://ha.privateeyepi.com/store/index.php?route=product/product&product_id=84&search=dht22.

With all of that in hand, I thus wired up my Raspberry Pi 3 and sensor as shown in Figure 2-10.

Figure 2-10. *Wiring up the DHT11 sensor to the Raspberry Pi*

To wire it up correctly, I followed the instructions included with the SunFounder kit for the DHT11 sensor as well as the appropriate pin mappings for the Raspberry Pi here:

https://ms-iot.github.io/content/en-US/win10/samples/PinMappingsRPi2.htm

Per the instructions for the DHT11 sensor that came with the SunFounder kit, the right pin is ground, the middle pin is power, and the left pin is Signal, or data. Thus, following the pin mappings from the link above and the instructions from SunFounder kit, the brown wire (my ground wire) goes from the right pin on the sensor to pin 6 on the Raspberry Pi. The red wire, my power wire, goes from the middle pin on the sensor to pin 2 (5 volt power) on the Pi, and the green wire, my data wire, goes from the left pin on the sensor to pin 4 (GPIO) on the Pi.

Now it's time to write the code to tell the Pi and the sensor to do their thing. Do the following:

1. Open Visual Studio 2015 and go to **File ➤ New ➤ Project**. From the Templates select **Windows ➤ Universal ➤ Blank app (Universal Windows)**. Select a location to save the project to and click the **Ok** button.

2. Download the source code for the DHT sensor from https://github.com/porrey/dht/tree/master/source/Windows%2010%20IoT%20Core/DHT%20Solution/Sensors.Dht and add the Sensors.DHT project to your solution.

3. Optionally, the DHT libraries were recently added to Nuget.org as a Nuget package. Thus, you can add the DHT sensor libraries by running command *Install-Package Dht* in the Package Manager Console window.

4. Add a reference to the Sensors.DHT project to the project you just created.

5. Next, add a reference to your project for Windows 10 IoT Extensions for UWP. To do this, right-click the references node in the Solution Explorer and select Add. From the left-hand tree, select **Universal Windows ➤ Extensions** and then select **Windows 10 IoT Extensions for UWP** and click the **Ok** button.

6. Next, using Nuget, add the **Microsoft.Azure.Devices.Client** package to your project.

7.   At the top of your MainPage.xaml.cs file, add the following imports:

```
using Sensors.Dht;
using Microsoft.Azure.Devices.Client;
using Windows.Devices.Gpio;
using System.Text;
```

8.   Add the following constants and variables:

```
private const int DHTPIN = 4;
private IDht dht = null;
private GpioPin dhtPin = null;
private DispatcherTimer sensorTimer = new DispatcherTimer();
```

9.   In the MainPage() constructor add the following code:

```
dhtPin = GpioController.GetDefault().OpenPin(DHTPIN, GpioSharingMode.Exclusive);
dht = new Dth11(dhtPin, GpioPinDriveMode.Input);
sensorTimer.Interval = TimeSpan.FromSeconds(1);
sensorTimer.Tick += sensorTimer_Tick;
sensorTimer.Start();
```

10.   In the class add the following methods:

```
private void sensorTimer_Tick(object sender, object e)
{
            readSensor();
}
private async void readSensor()
{
    DhtReading reading = await dht.GetReadingAsync().AsTask();
    if (reading.IsValid)
    {
        double temp = ConvertTemp.ConvertCelsiusToFahrenheit(reading.Temperature);
        string message = "{\"temperature\" :" + temp.ToString() + ", \"humidity\":" +
        reading.Humidity.ToString() + "}";
        listBox.Items.Add(message.ToString());
    }
}
```

11.   Next, add the following class:

```
static class ConvertTemp
{
    public static double ConvertCelsiusToFahrenheit(double c)
    {
        return ((9.0 / 5.0) * c) + 32;
    }
}
```

12.   Now open the MainPage and add a ListBox to the form. This is simply to visual
       see that data is being returned when deployed to the Pi.

13. Next, you need to deploy the solution to your Raspberry Pi. In Visual Studio, from the tool bar, set the solution platform to **ARM** and the target as **Remote Machine,** as shown in Figure 2-11.

Figure 2-11. *Running the application to publish to the Raspberry Pi*

14. Open the project properties by double-clicking the **Properties** node in Solution Explorer and navigate to the **debug** settings.

15. In the **Remote machine:** textbox you need to enter the IP address/name of your Raspberry Pi. To do this, click the Find button and wait a few moments for Visual Studio to search for your devices. Your Raspberry Pi should show in the Auto Detected section. Click your Raspberry Pi and click the **Select** button. If your device doesn't show, it should be called **minwinpc** so enter this value.

16. Press the **green play button** next Remote Machine or press **F5** to build and deploy the solution to your Raspberry Pi. Your device should now start sending temperature and humidity readings, as shown in Figure 2-12.

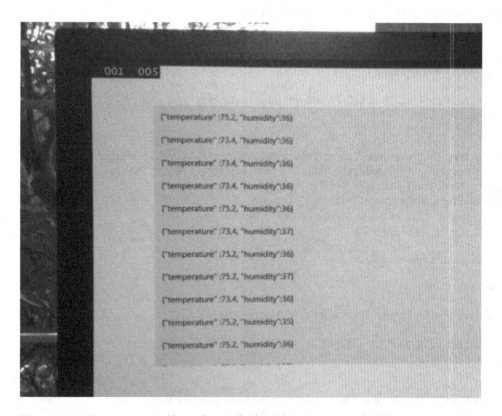

Figure 2-12. *Temperature and humidity results*

By default, the DHT11 sensor returns the temperature in Celsius. Thus, the method to convert it to Fahrenheit. If you are ok with Celsius, comment out the two lines directly above the `listbox.Items.Add` call, and then uncomment out the first `string message` line.

If you are using the DHT11 sensor, the code above will also work. All you need to do is change the following line to reference the DHT22 instead of the DHT11:

```
dht = new Dht22(dhtPin, GpioPinDriveMode.Input);
```

I'll also point out the line of code used above:

```
private const int DHTPIN = 4;
```

Notice that the DHT pin is set to 4. As mentioned above, the green wire, my data wire, goes from the left pin on the sensor to pin 4 (GPIO) on the Pi so I am specifying which pin to get the data from.

At this point in the example, installing the Nuget package `Microsoft.Azure.Devices.Client` wasn't necessary but it will come in handy in an upcoming chapter with this sample so let's get it out of the way. This package provides device support for sending messages to an Azure IoT Hub, which will be discussed in the next chapter.

To get the temperature to fluctuate to simulate a malfunction (either the sensor going bad or the actual temperature not remaining consistent), I took a hair dryer and an air canister and rotated blowing air on the sensor to get the temperature to go up and down. Now, truth be told, for this example you didn't need a breadboard. To make things simpler, you could have simply connected the sensor via the wires directly to the Raspberry Pi using the same pin configuration.

For more information on the Raspberry Pi, visit `www.raspberrypi.org/`.

Before moving on to the Tessel, the next section will discuss a device that is branded as the "ultimate topping for your Pi!"

FEZ HAT

The previous section used a breadboard, wires, and a DHT22 sensor to generate data via the Raspberry Pi. This meant there was a need to understand the pin layout on the devices to wire up the sensor correctly. This comes in handy if you need to wire up multiple sensors, but there is an easier way via the FEZ HAT, shown in Figure 2-13.

Figure 2-13. *The FEZ HAT*

The FEZ (Fast and Easy) HAT is a simple device that simply sits on top of the Raspberry Pi, as shown in Figure 2-14, and includes a wide variety of features:

- Temperature Sensor
- Light Sensor
- Accelerometer
- Two Server Motor Connections

For a complete list of features and functionality, visit www.ghielectronics.com/catalog/product/500.

Figure 2-14. *FEZ HAT with a Raspberry Pi*

Using the FEZ HAT is quite simple. The first step is to install the drivers from Nuget. While the code will be very similar, go ahead and create a new blank UWP app and run the following commands in the Package Manager Console to install the appropriate FEZ HAT drivers:

```
Install-Package GHIElectronics.UWP.Shields.FEZHAT
```

Once the package is installed, add the following statement to the MainPage form in the using section:

```
using GHIElectronics.UWP.Shields;
using Newtonsoft.Json;
```

Before you start adding code, open up the MainPage and add a ListBox to the form, similar to what you did in the previous section. And, just like you did in the previous example, you will need to add a reference to your project for Windows 10 IoT Extensions for UWP. To do this, right-click the references node in the Solution Explorer and select Add. From the left-hand tree, select **Universal Windows ➤ Extensions** and then select **Windows 10 IoT Extensions for UWP** and click the **Ok** button.

Next, in the code behind the MainPage, add the following statements above the MainPage() constructor:

```
private FEZHAT hat;
private DispatcherTimer timer;
private bool next;
```

Add the following to the MainPage() constructor:

```
this.SetupHat();
```

Next, add the SetupHat() method:

```
private async void SetupHat()
{
    this.hat = await FEZHAT.CreateAsync();
    this.timer = new DispatcherTimer();
    this.timer.Interval = TimeSpan.FromSeconds(1);
    this.timer.Tick += this.Timer_Tick;
    this.timer.Start();
}
```

Next, add the Time_Tick() method:

```
private void Timer_Tick(object sender, object e)
{
    readSensor();
}
```

Lastly, add the readSensor() method:

```
private async void readSensor()
{

    var light = this.hat.GetLightLevel();

    // Temperature Sensor
    var temp = this.hat.GetTemperature();

    double ftemp = ConvertTemp.ConvertCelsiusToFahrenheit(temp);

    System.Diagnostics.Debug.WriteLine("Temperature: {0} °C, Light {1}", ftemp.
ToString("N2"), light.ToString("N2"));

    string message = "{\"temperature\" :" + temp.ToString() + "}";

    //send to listbox
    listBox.Items.Add(message);
}
```

Similarly, add the ConvertTemp class so that the temperature returned from the FEZ HAT can be converted to Fahrenheit.

The entire code listing is below for your reference:

```
using System;
using System.Collections.Generic;
using System.IO;
using System.Linq;
```

```
using System.Runtime.InteropServices.WindowsRuntime;
using Windows.Foundation;
using Windows.Foundation.Collections;
using Windows.UI.Xaml;
using Windows.UI.Xaml.Controls;
using Windows.UI.Xaml.Controls.Primitives;
using Windows.UI.Xaml.Data;
using Windows.UI.Xaml.Input;
using Windows.UI.Xaml.Media;
using Windows.UI.Xaml.Navigation;
using GHIElectronics.UWP.Shields;

// The Blank Page item template is documented at http://go.microsoft.com/fwlink/?LinkId=402
352&clcid=0x409

namespace App1
{
    /// <summary>
    /// An empty page that can be used on its own or navigated to within a Frame.
    /// </summary>
    public sealed partial class MainPage : Page
    {
        private FEZHAT hat;
        private DispatcherTimer timer;
        private bool next;

        public MainPage()
        {
            this.InitializeComponent();
            this.SetupHat();
        }

        private async void SetupHat()
        {
            this.hat = await FEZHAT.CreateAsync();
            this.timer = new DispatcherTimer();
            this.timer.Interval = TimeSpan.FromSeconds(1);
            this.timer.Tick += this.Timer_Tick;
            this.timer.Start();
        }

        private void Timer_Tick(object sender, object e)
        {
            readSensor();
        }

        private async void readSensor()
        {

            var light = this.hat.GetLightLevel();
```

```
            // Temperature Sensor
            var temp = this.hat.GetTemperature();

            double ftemp = ConvertTemp.ConvertCelsiusToFahrenheit(temp);

            System.Diagnostics.Debug.WriteLine("Temperature: {0} °C, Light {1}", ftemp.
            ToString("N2"), light.ToString("N2"));

            string message = "{\"temperature\" :" + temp.ToString() + "}";

            //send to listbox
            listBox.Items.Add(message);
        }

    }
    static class ConvertTemp
    {
        public static double ConvertCelsiusToFahrenheit(double c)
        {
            return ((9.0 / 5.0) * c) + 32;
        }
    }
}
```

This code sends the code to the ListBox. This is a good way to make sure your FEZ HAT is working. Once working, comment out the three lines of code below the //send to Azure comment and rerun it. The code above also gets the light sensor information but does not display it. Feel free to change the code to display the light data.

To run this application, you will need to follow the same simple steps for adding the IP address or name of the Raspberry Pi to the project. Simply click the Properties node in Solution Explorer and go to the debug settings. Enter the IP address and click Select. You should be good to go. Make sure your output is Remote Machine, as shown in Figure 2-11, and then compile and run the project! You should see the temperature displayed in the ListBox each second.

In Chapter 4, you'll modify these two examples to have the data saved to Azure IoT Hub. But first, let's do this same thing using a Tessel.

Tessel

The Tessel board is an open source, community-driven project driven by the Tessel Project. The aim of the Tessel is the development and learning for IoT devices. The recently released Tessel 2, shown in Figure 2-15, shipped in March of 2016. Compared to the Raspberry Pi, it is a bit smaller and a bit less powerful, but it is still very functional and is a fantastic option for IoT development. In Figure 2-15, the device on the left is the climate module which will be used to gather temperature data, similar to what was done with the Raspberry Pi.

Figure 2-15. *Tessel 2 with a climate sensor*

New to Tessel 2 is improved Wi-Fi, an Ethernet jack, two USB ports, and support for Node.js (whereas Tessel 1 was JavaScript-based). Tessel uses Node.js as its language of choice but should support Python and Rust in the future.

Let's get started. For these examples you can use any code editor. Two popular editors are Visual Studio Code (`https://code.visualstudio.com/`) or Notepad++ (`https://notepad-plus-plus.org/`). I'll be using Visual Studio Code.

Before plugging in the Tessel, download the following:

- Download version 4.4.4 of Node.js from `https://nodejs.org/download/release/v4.4.4/`

- Download the Zadig USB driver install from `http://zadig.akeo.ie/`

Once these are downloaded, run the Node.js install. You probably don't need to reboot after the installation completes, but it might be a good idea just in case. Next, plug in your Tessel. Your Tessel came with a USB power cable. Go ahead and connect your Tessel to power it on. A blue light will appear on your Tessel.

Next, run the Zadig USB tool, which will install the generic USB drivers such as WinUSB. When the Zadig tool launches, accept all the defaults and simply click the Install button. The installation will only take a few seconds. Once installed, close the Zadig tool.

Open a Command prompt as Administrator and at the command prompt type the following to install the Tessel drivers:

```
npm install i -g t2-cli
```

The output will look similar to Figure 2-16.

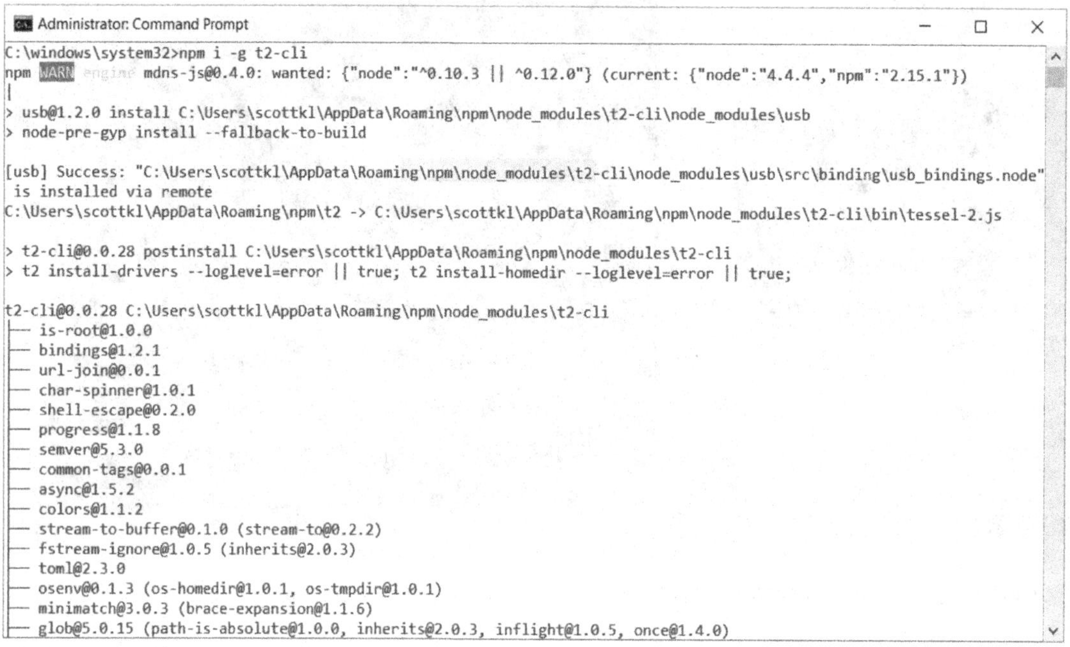

```
Administrator: Command Prompt                                                    —    □    ×
C:\windows\system32>npm i -g t2-cli
npm WARN engine  mdns-js@0.4.0: wanted: {"node":"^0.10.3 || ^0.12.0"} (current: {"node":"4.4.4","npm":"2.15.1"})
|
> usb@1.2.0 install C:\Users\scottkl\AppData\Roaming\npm\node_modules\t2-cli\node_modules\usb
> node-pre-gyp install --fallback-to-build

[usb] Success: "C:\Users\scottkl\AppData\Roaming\npm\node_modules\t2-cli\node_modules\usb\src\binding\usb_bindings.node"
 is installed via remote
C:\Users\scottkl\AppData\Roaming\npm\t2 -> C:\Users\scottkl\AppData\Roaming\npm\node_modules\t2-cli\bin\tessel-2.js

> t2-cli@0.0.28 postinstall C:\Users\scottkl\AppData\Roaming\npm\node_modules\t2-cli
> t2 install-drivers --loglevel=error || true; t2 install-homedir --loglevel=error || true;

t2-cli@0.0.28 C:\Users\scottkl\AppData\Roaming\npm\node_modules\t2-cli
├── is-root@1.0.0
├── bindings@1.2.1
├── url-join@0.0.1
├── char-spinner@1.0.1
├── shell-escape@0.2.0
├── progress@1.1.8
├── semver@5.3.0
├── common-tags@0.0.1
├── async@1.5.2
├── colors@1.1.2
├── stream-to-buffer@0.1.0 (stream-to@0.2.2)
├── fstream-ignore@1.0.5 (inherits@2.0.3)
├── toml@2.3.0
├── osenv@0.1.3 (os-homedir@1.0.1, os-tmpdir@1.0.1)
├── minimatch@3.0.3 (brace-expansion@1.1.6)
└── glob@5.0.15 (path-is-absolute@1.0.0, inherits@2.0.3, inflight@1.0.5, once@1.4.0)
```

Figure 2-16. *Setting up the Tessel environment*

Once the Tessel drivers are installed, type the following command and press Enter. This command will list your Tessel and any other Tessels available over Wi-Fi and USB.

```
t2 list
```

As shown in Figure 2-17, your Tessel should be listed, specifying that is connected via USB. Next, type the following command and press Enter:

```
t2 provision
```

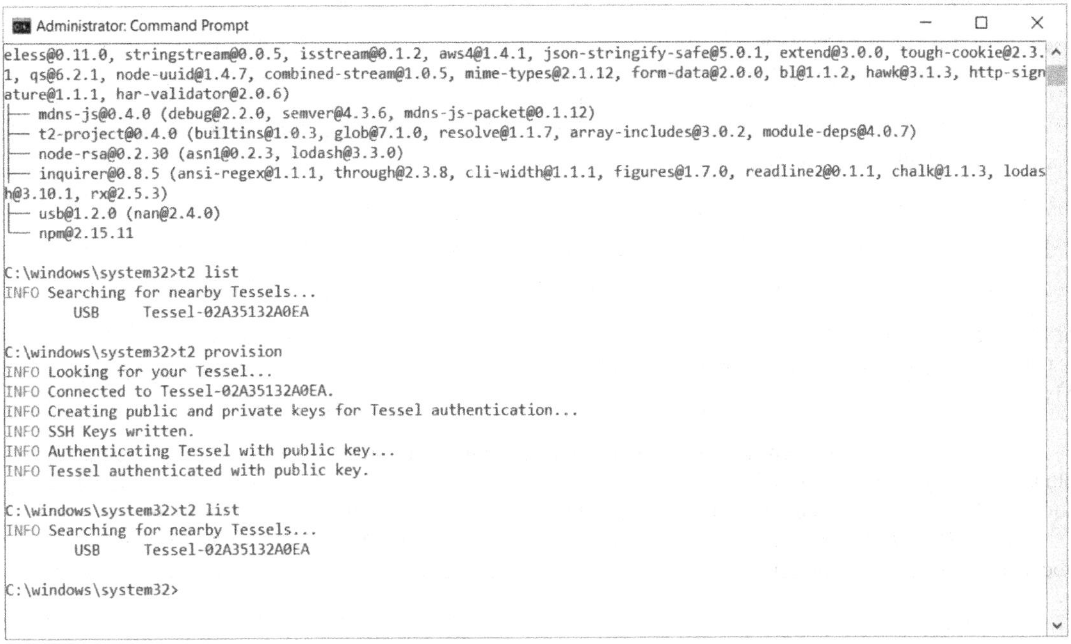

```
Administrator: Command Prompt                                            —    □    ×
eless@0.11.0, stringstream@0.0.5, isstream@0.1.2, aws4@1.4.1, json-stringify-safe@5.0.1, extend@3.0.0, tough-cookie@2.3.
1, qs@6.2.1, node-uuid@1.4.7, combined-stream@1.0.5, mime-types@2.1.12, form-data@2.0.0, bl@1.1.2, hawk@3.1.3, http-sign
ature@1.1.1, har-validator@2.0.6)
├── mdns-js@0.4.0 (debug@2.2.0, semver@4.3.6, mdns-js-packet@0.1.12)
├── t2-project@0.4.0 (builtins@1.0.3, glob@7.1.0, resolve@1.1.7, array-includes@3.0.2, module-deps@4.0.7)
├── node-rsa@0.2.30 (asn1@0.2.3, lodash@3.3.0)
├── inquirer@0.8.5 (ansi-regex@1.1.1, through@2.3.8, cli-width@1.1.1, figures@1.7.0, readline2@0.1.1, chalk@1.1.3, lodas
h@3.10.1, rx@2.5.3)
├── usb@1.2.0 (nan@2.4.0)
└── npm@2.15.11

C:\windows\system32>t2 list
INFO Searching for nearby Tessels...
        USB     Tessel-02A35132A0EA

C:\windows\system32>t2 provision
INFO Looking for your Tessel...
INFO Connected to Tessel-02A35132A0EA.
INFO Creating public and private keys for Tessel authentication...
INFO SSH Keys written.
INFO Authenticating Tessel with public key...
INFO Tessel authenticated with public key.

C:\windows\system32>t2 list
INFO Searching for nearby Tessels...
        USB     Tessel-02A35132A0EA

C:\windows\system32>
```

Figure 2-17. *Configuring the Tessel 2*

The provision command authorizes your computer to access and push code to the connected Tessel 2 via SSH. As shown in Figure 2-17, the provisioning generates public and private keys for Tessel authentication.

If you don't like the name of your Tessel, you can rename your Tessel by typing t2 rename and specifying a new name. For example,

```
t2 rename "newname"
```

Executing t2 update updates the firmware of the Tessel, and executing npm install i -g t2-cli any time will check for, and update, the Tessel drivers.

It's time to do some actual work. Power off the Tessel and then plug in the climate sensor to Port A of the Tessel with the electrical components up. Plug the Tessel back in. Create a directory on your hard drive, and in the command prompt, navigate to that directory. Install the climate sensor drivers by executing the following command:

```
npm install climate-si7010
```

Next, open up Visual Studio Code or your favorite text editor and enter the following code:

```
var tessel = require('tessel');
var climatelib = require('climate-si7020');

var climate = climatelib.use(tessel.port['A']);

climate.on('ready', function () {
  console.log('Connected to climate module');
```

```
  // Loop forever
  setImmediate(function loop () {
    climate.readTemperature('f', function (err, temp) {
      climate.readHumidity(function (err, humid) {
        console.log('Degrees:', temp.toFixed(4) + 'F', 'Humidity:', humid.toFixed(4) + '%RH');
        setTimeout(loop, 300);
        });
      });
    });
});

climate.on('error', function(err) {
  console.log('error connecting module', err);
});
```

The code above loops each second and gets the temperature and humidity reading from the sensor plugged into the Tessel. I have a few things to point out. The first three lines of code specify that the following code needs the Tessel and climate drivers, and that the climate sensor is plugged in to Port A on the Tessel.

Save the file as climate.js, and in the same command prompt used earlier to the directory you created above, type the following and press Enter:

```
t2 run climate.js
```

The t2 run command finds your connected Tessel, compiles the code (in this case, climate.js), and then deploys the code to the Tessel and runs it, as shown in Figure 2-18.

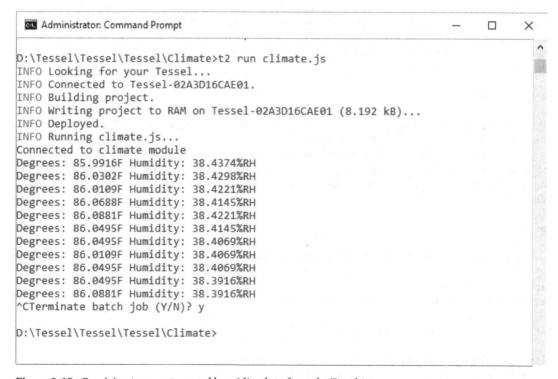

Figure 2-18. *Receiving temperature and humidity data from the Tessel 2*

Congratulations on getting your Tessel up and running. Along with the previous two examples, Chapter 4 will also modify this example so that the data is sent to Azure.

For more information on the Tessel, visit `https://tessel.io/`.

Big data and the IoT are about gathering, processing, analyzing, and gaining valuable insights into extremely large quantities of data. Remember that two of the three Vs are volume and velocity. Thus, imagine not one but hundreds or thousands of these boards generating data.

Summary

This chapter focused on two different devices that generate data, the Raspberry Pi and Tessel. First, I walked you through how to install Windows 10 IoT Core on a Pi 2 and a Pi 3, and then I walked you through an example of using the Pi and a temperature sensor to generate data. That was followed by a similar example using a Tessel board to show Microsoft's support for open source technologies.

The next chapter will introduce the Microsoft Azure IoT Hub and its importance in the IoT service spectrum.

PART II

Data on the Move

CHAPTER 3

■ ■ ■

Azure IoT Hub

In the world today there are millions of devices of different types generating an enormous amount of data. A large number of these devices are part of a larger IoT solution in which the devices are sending their data to the cloud for storage, processing, and analysis. At a high level, IoT solutions can be broken down into core essentials of *device connectivity* and *data processing and analysis*, as shown in Figure 3-1.

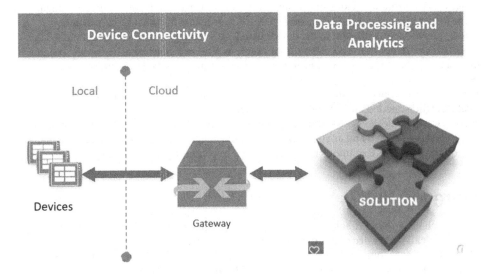

Figure 3-1. IoT solution core essentials

Device connectivity is simply devices that generate and collect data which is then sent to a cloud gateway. The cloud gateway acts as a mediator that gathers the incoming data and makes it available for further processing by other services and processes of the IoT solution. Yet, within the IoT solution architecture there are real and distinct challenges that exist. These challenges come in the form of how to create a secure and dependable connection between the devices and the back-end solution.

Think about how devices are being used today. Devices have been connected to cows, cars, and spaceships. Your phone, watch, step counters, and fitness trackers are data-generating devices. So, when dealing with device connectivity, the challenge is to figure out the best and most efficient method and approach for enabling and providing not only securely connected devices but also reliably connected devices.

© Scott Klein 2017

S. Klein, *IoT Solutions in Microsoft's Azure IoT Suite*, DOI 10.1007/978-1-4842-2143-3_3

IoT solutions are not your typical client app. Think about it for a minute. As mentioned above, these IoT devices aren't sitting on your desktop and you don't hit them with your browser. No, these devices are out in pastures, on the street, or orbiting the earth. They are in farms and factories, appliances and gadgets.

As such, special attention needs to be paid to their unique, perplexing characteristics, such as

- Slow or unreliable network connectivity

- Limited power resources

- Lack of physical access to the device

- Possible use of proprietary or custom application protocols

- Human device interaction

While this list is in no way complete, it should give you an idea of the real challenges faced by IoT solutions today. In addition, there is another key characteristic of devices that cannot be overlooked. It should be noted that the arrow between the devices and the gateway (as well as the gateway and the IoT Solution) in Figure 3-1 is bidirectional. This is due to the fact that IoT solutions require secure, *bidirectional* communication between devices and cloud gateways.

Thus, not only are devices sending data (device-to-cloud communication), but they can also receive and process messages and information (cloud-to-device communication) from a cloud endpoint. Imagine a scenario where the IoT solution might send a message to the device telling it to change configuration values. For example, a message could be sent to the device to tell it to change the rate at which it pulls data or change the upper or lower temperature alert limits. For example, iPads for years have been able to signal back and wipe themselves if they are stolen. In another example, a company in the UK is limiting water based on telemetry from the sensors on industrial water meters.

What is needed is a service that overcomes these challenges, not only providing proper security and reliability but scalability when needed. Introducing Azure IoT Hub.

In the previous chapter, the devices generated temperature and humidity data but the data was never stored or analyzed. How data is stored and analyzed will be covered in later chapters; this chapter will focus on the needs and requirements of IoT solutions that are pertinent to communicating with devices.

What Is Azure IoT Hub?

To overcome the innate challenges of IoT solutions discussed earlier, Microsoft introduced Azure IoT Hub, a fully managed communications service that provides highly secure, dependable, and scalable messaging between IoT devices and IoT solutions. Azure IoT Hub can be viewed as a high-scale gateway that acts as an enabler and manager of all bidirectional communication to and from devices, as shown in Figure 3-2.

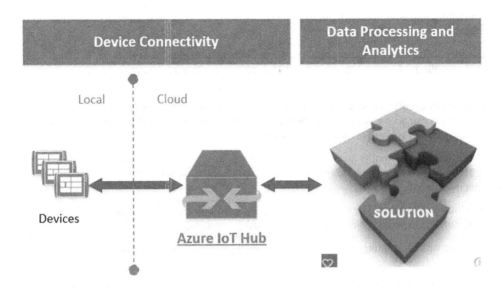

Figure 3-2. The role of IoT Hub

At its core, Azure IoT Hub is designed to overcome the device connection limitations and challenges inherent to IoT solutions discussed earlier, including

- Reliable and dependable device-to-cloud and cloud-to-device messaging
- Real-time device registry
- Secure communication via per-device security credentials and access control
- Extensive device connectivity monitoring
- Event monitoring
- Libraries and SDKs available for most languages

Let's explore these benefits in a little more detail.

Why Use Azure IoT Hub?

While there are already existing Azure services that provide device-to-cloud messaging, the benefits of Azure IoT Hub provide the key, essential necessities needed in IoT solutions. The architectural goals and principles that went in to designing and building IoT Hub were founded on four vital areas:

- Support for both hardware and software scenarios, such as the wide collection of devices, environments, and scenarios
- Security should be the foundation of the service across all aspects, including data protection, and device and user identity
- Support for millions of simultaneous connected devices
- Composability to allow for the extension of various components

With these principles and goals in mind, the benefits of Azure IoT hub can be summarized into the following:

- **Scalability**: The ability to support millions of simultaneous connected devices as well as millions of events per second. IoT Hub automatically scales as you add devices.

- **Per-Device Authentication and Security**: Complete, fine-grained control over which devices can access your IoT solution as well as ensuring that any cloud-to-device commands are sent to the correct device.

- **Device Monitoring**: Identify device connectivity issues through detailed operation logs. These logs contain device identity management operations and device connectivity events.

- **Extensibility**: IoT Hub can be extended to provide support for custom protocols. Extensibility through the use of first-party and third-party technologies.

- Support for various languages and platforms through IoT device SDKs and device libraries.

With these benefits as a backdrop to this chapter, an architectural discussion into data flow and communication will be helpful in creating and configuring the IoT Hub later in this chapter and for use in the next chapter.

Architectural Overview

A key element, and one of the challenges discussed earlier, is how IoT devices can connect to IoT solutions. For Azure IoT Hub, devices can connect either directly or indirectly. Figure 3-3 shows the logical communication flow from the different types of devices into the IoT Hub.

IP-capable devices can directly connect using IoT Hub-supported protocols like HTTP, MQTT, AMQPS, and TLS. Devices that do not support these protocols can still connect to Azure IoT Hub via a custom cloud protocol gateway. Those devices that cannot access the Internet directly, such as devices using industry-specific protocols that work on internal networks, can still benefit from IoT Hub by using what is called a field gateway.

AMQP stands for Advanced Message Queueing Protocol and is an open standard application layer protocol finish. MQTT is a machine-to-machine connectivity protocol designed as a lightweight publish/subscribe messaging transport, which makes it very suitable for use with IoT.

A protocol gateway for Azure IoT is simply an open-source framework that provides bidirectional communication between devices. As an open-source framework it supports custom gateways and protocol adaptions, including MQTT.

A field gateway acts as a communication enabler, and is typically an appliance or general purpose software, or Internet-capable device that relays messages from smaller non-Internet-capable devices. They differ from traffic routers in that they manage information and access flow.

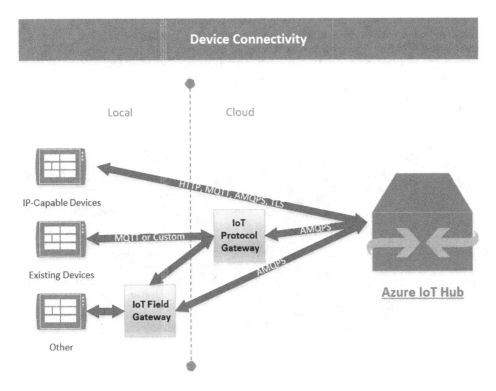

Figure 3-3. *IoT Hub device connectivity*

The architecture described and implemented by Azure IoT Hub enables two specific communication patterns, which have been mentioned, but it's worth spending a bit more time on them.

- **Device-to-cloud**: As mentioned, device-to-cloud communication is refers to data being sent from the device to the cloud. For example, from a Raspberry Pi to Azure IoT Hub.

- **Cloud-to-device**: On the flip side, IoT solutions can use IoT Hub to send messages to individual devices.

Two concepts that frequently come up when talking about device-to-cloud communication are *hot path* and *cold path* processes. Understanding these concepts will help determine the optimum settings for device-to-cloud communications. At a high level, hot path refers to data that needs immediate or near-immediate processing and analysis. For example, as data comes in to IoT Hub, it is immediately picked up by Azure Stream Analytics for real-time analysis. Cold path refers to data stored to be processed later. For example, you might be gathering data that you want to store in a relational database for future processing or analysis.

Creating an IoT Hub

Now that you have a good foundation and understanding of IoT Hub, this section will walk you through creating and configuring an Azure IoT Hub. Here you will create and configure the IoT Hub. In the next chapter, you will hook up the IoT Hub to your Raspberry Pi.

As mentioned in the introduction, all the examples in this book, beginning with this one and throughout the rest of the book, will require a Microsoft Azure subscription. You can create a free, 30-day trial Azure account by visiting https://azure.microsoft.com/en-us/free/. If you have a Microsoft MSDN subscription, you can activate Microsoft Azure. Details can be found on your individual MSDN account site.

OK, let's get started. Open your favorite browser and navigate to http://portal.azure.com/. When prompted, log in. Once in the portal, click New, select Internet of Things, and then select Azure IoT Hub, as shown in Figure 3-4.

Figure 3-4. *Creating an Azure IoT Hub from the Azure Portal*

The IoT Hub blade will appear, similar to the one in Figure 3-5. In the blade, enter or select the following:

- **Name**: The name of your IoT Hub. Since I am connecting my Raspberry Pi 3 to the hub, I named my hub MyPi3Hub.

- **Pricing and Scale Tier**: Click that section of the blade and select the Free F1 tier.

- **Resource Group**: Either select an existing Resource Group or create a new one.

- **Location**: Select the appropriate region in which to host the IoT Hub.

By default, the Pricing and Scale Tier will default to a Standard S1 tier. Clicking in that section of the blade will show that there are three tiers; Free, S1, and S2. The difference between the tiers is the number of messages sent to and from IoT Hub per unit each day. For the purposes of this demo and the book, the Free tier will suffice. Only one Free tier can be created per subscription.

Both the IoT Hub units and device-to-cloud partitions will have default values (one for IoT Hub units and four for device-to-cloud partitions). Leave those values as is, but I'll describe what they are.

IoT Hub units define the performance of your IoT Hub. For example, the free tier provides 8,000 messages per unit, per day. However, with the free tier, you are only allowed a maximum of one unit. The standard S1 and S2 tiers provide up to 200 units. Thus, if you need more performance, increase your number of units within your IoT Hub.

When creating an IoT Hub, your performance sweet spot will be determined by the number of devices connecting to IoT Hub along with how many messages will be processed by IoT Hub. Doing that math will give you an idea of what tier you need, along with the number of units within that tier. You are essentially billed by the number of units in your IoT Hub.

IoT Hub provides message streaming through a partitioned consumer pattern, where a partition is simply an ordered sequence of events. As new messages come in, they are added to the end of the sequence. An IoT Hub can have multiple partitions, each operating independent of other partitions and each containing its own sequence of data, thus growing at different rates.

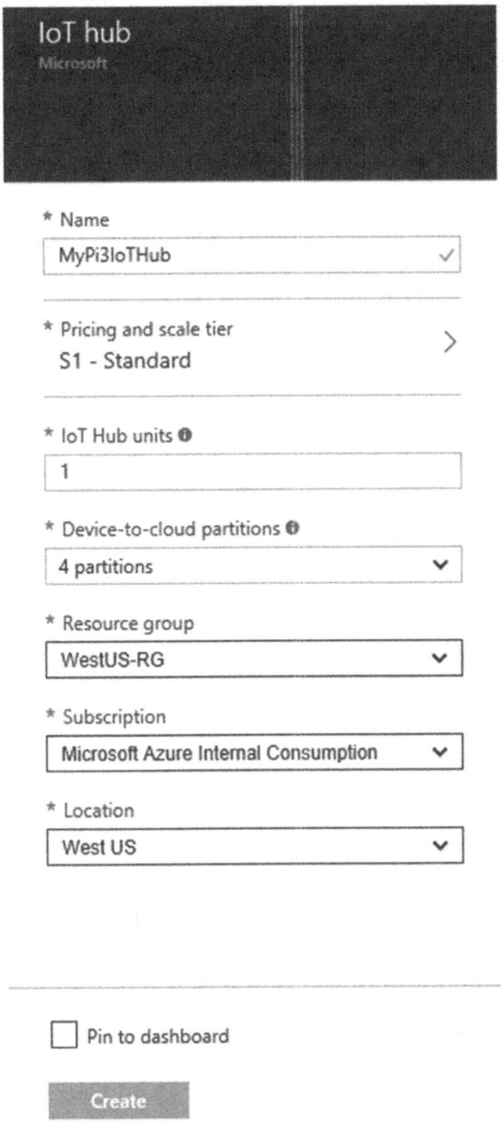

Figure 3-5. *IoT Hub creation settings*

Click the "Pin to dashboard" checkbox and then click Create. The creation of the IoT Hub will take a couple of minutes. Once it is created, you will see a new icon on your dashboard. Click the new IoT Hub icon on your dashboard to open up both the Essentials blade and the Settings blade for the newly created IoT Hub, similar to Figure 3-6.

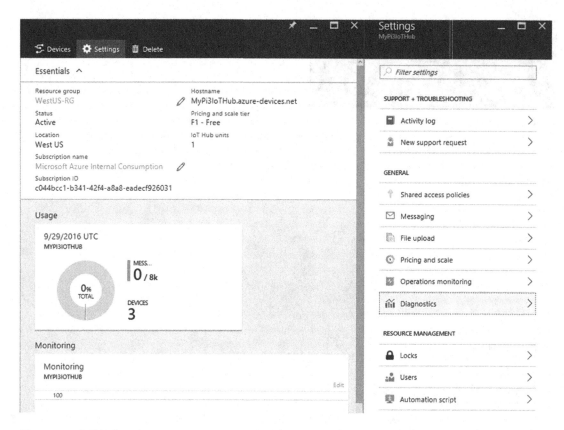

Figure 3-6. *IoT Hub settings*

The left Essentials blade provides a summary of the IoT Hub as well as usage information such as the number of messages processed by the IoT Hub and the number of devices registered. Also included in this is monitoring information which shows all event and connectivity information being collected.

The right Settings blade is where further configuration of items such as messaging, diagnostics, and operations monitoring takes place. The following section will discuss these settings.

Messaging Settings

The Messaging blade contains configuration settings pertaining to cloud-to-device and device-to-cloud message communication, as shown in Figure 3-7.

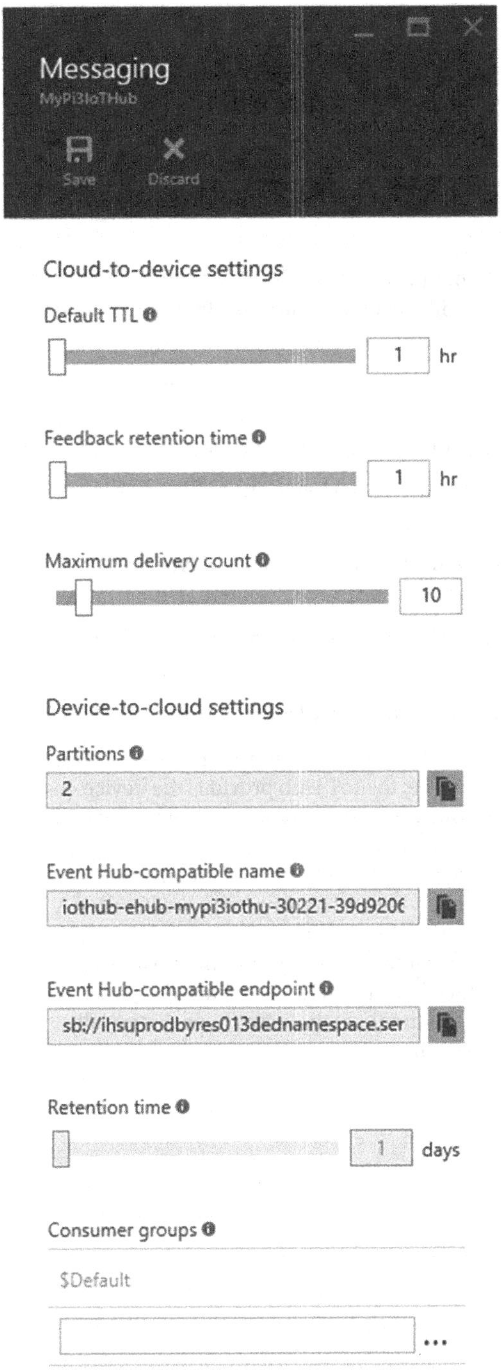

Messaging
MyPi3IoTHub

Save   Discard

Cloud-to-device settings

Default TTL ❶

| 1 | hr

Feedback retention time ❶

| 1 | hr

Maximum delivery count ❶

| 10 |

Device-to-cloud settings

Partitions ❶

| 2 |

Event Hub-compatible name ❶

| iothub-ehub-mypi3iothu-30221-39d920€ |

Event Hub-compatible endpoint ❶

| sb://ihsuprodbyres013dednamespace.ser |

Retention time ❶

| 1 | days

Consumer groups ❶

$Default

| | ... |

Figure 3-7. *IoT Hub messaging settings*

Most of the settings come with default configuration values but I'll explain these settings.

Cloud-To-Device Settings

- **Default TTL (Time-to-Live)**: Specifies the amount of time (hours) a message is available for the device to consume before it is expired by IoT Hub. Valid values are from 1 to 48.

- **Feedback Retention Time**: Specifies how long in hours the IoT Hub will maintain the feedback for expiration or delivery of cloud-to-device messages. Valid values are from 1 to 48.

- **Maximum Delivery Count**: Specifies the number of times the IoT Hub will attempt to deliver a cloud-to-device message to a device. Valid values are from 1 to 100.

Device-To-Cloud Settings

- **Partitions**: Specifies the number of partitions for your device-to-cloud events.

- **Event Hub-Compatible Name**: Specifies the Event Hub name when using SDKs or other integrations that expect to read from Event Hub.

- **Event Hub-Compatible Endpoint**: Specifies the Event Hub endpoint when using SDKs or other integrations that expect to read from Event Hub.

- **Retention Time**: Specifies how long in days this IoT Hub will maintain device-to-cloud events.

- **Consumer Groups**: Used by applications to pull data from the IoT Hub.

Any time an IoT Hub is created, an Event Hub is also created internally. It is common to use both an Azure IoT Hub and Event Hub in the same IoT solution. For example, the IoT Hub provides the device-to-cloud communication with the Event Hub picking up the real-time processing later in the data processing stages. Thus, it is useful to have access to the Event Hub in some situations.

Consumer groups have been mentioned a few times in this chapter already, and will be mentioned a few more times throughout the book, so a few words on *consumers* and *consumer groups* are in order. A *consumer* is simply any entity (application or process) that reads data from hub (Event Hub or IoT Hub). A *consumer group* is a view of a Hub (Event Hub or IoT Hub) as a whole, including state and position. Consumer groups provide consuming applications, such as Azure Stream Analytics, to have their own independent and separate view into the event stream and read the stream at their own pace. Essentially, consumer groups are used by applications to pull data from the Hub.

Event Hub will be covered in a later chapter, but for now suffice it to say that both are event processing services that enable event and telemetry ingress to the cloud, and both allow you to create 20 consumer groups. However, the main differences are the following:

- IoT Hub provides device-to-cloud and cloud-to-device messaging, whereas Event Hub only supports event ingress (device-to-cloud).

- Both support AMQP and AMQP over WebSockets and HTTP/1, but IoT Hub also works with the IoT Hub Protocol Gateway, which supports custom protocols.

- Both use per-device identity control for secure communication.

- IoT Hub simultaneously supports millions of connected devices, where Event Hub supports a more limited number of connections.

A bit more needs to be said on the custom protocols because this is quite a big topic, one that probably warrants a whole chapter or two by itself. However, for the purposes of this chapter and to build on the custom protocols idea, the Microsoft Azure IoT Protocol Gateway needs some attention. The Microsoft Azure IoT Protocol Gateway is a framework that provides a programming model for building custom protocol adapters for a variety of protocols.

The Microsoft Azure IoT Protocol Gateway provides a number of configuration settings to define the behavior of the protocol, including support for Quality of Service (QoS), subscriptions and topics, message ordering, retry logic, authorization, encryption and decryption, and more. Usually there is no need to adjust the settings but you can if there is a need to adjust the protocols behavior.

You can read more about the Microsoft Azure IoT Protocol Gateway and download it at `https://github.com/Azure/azure-iot-protocol-gateway`.

Operations Monitoring Settings

The Operations Monitoring blade contains configuration settings pertaining to the type of information and events monitored by IoT Hub, as shown in Figure 3-8.

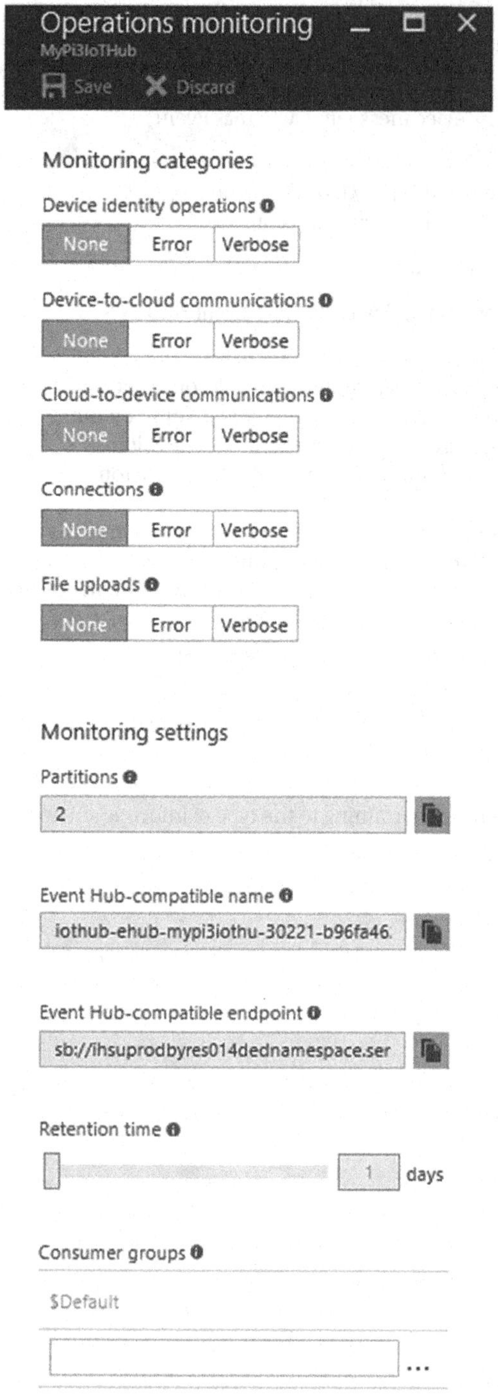

Operations monitoring — □ ✕
MyPi3IoTHub

🖫 Save   ✕ Discard

Monitoring categories

Device identity operations ❶

| None | Error | Verbose |

Device-to-cloud communications ❶

| None | Error | Verbose |

Cloud-to-device communications ❶

| None | Error | Verbose |

Connections ❶

| None | Error | Verbose |

File uploads ❶

| None | Error | Verbose |

Monitoring settings

Partitions ❶

2

Event Hub-compatible name ❶

iothub-ehub-mypi3iothu-30221-b96fa46.

Event Hub-compatible endpoint ❶

sb://ihsuprodbyres014dednamespace.ser

Retention time ❶

1 days

Consumer groups ❶

$Default

...

Figure 3-8. IoT Hub operations monitoring settings

Monitoring Categories

- **Device Identity Operations**: Logs errors related to operations on the device identity registry.

- **Device-to-Cloud Communications**: Logs errors related to device-to-cloud messaging.

- **Cloud-to-Device Communications**: Logs errors related to cloud-to-device messaging.

- **Connections**: Logs errors when a device connects or disconnects from the IoT Hub.

Each IoT Hub has a device identity registry that is available to use for creating per-device resources in the IoT Hub service. For example, it might be used as a queue to store up cloud-to-device messages to be sent to the device later.

When a user attempts to create, update, or delete an entry in the IoT Hub identity registry, the device identity operations track this information. This information comes in handy when you want to monitor certain scenarios such as device provisioning.

Monitoring Settings

- **Partitions**: Specifies the number of partitions for your operations monitoring events.

- **Event Hub-Compatible Name**: Specifies the Event Hub name when using SDKs or other integrations that expect to read from Event Hub.

- **Event Hub-Compatible Endpoint**: Specifies the Event Hub endpoint when using SDKs or other integrations that expect to read from Event Hub.

- **Retention Time**: Specifies how long in days this IoT Hub will maintain operation monitoring events.

- **Consumer Groups**: Used by applications to pull data from the IoT Hub.

Notice that the Event Hub settings are similar to the information on the Messaging blade. This is simply because one is for messaging and the other is for monitoring. Similar details are needed.

Diagnostics Settings

The Diagnostics blade contains configuration settings pertaining to diagnostics for your Azure IoT Hub. By default, diagnostics is turned off, which disables monitoring charts and alerts for your resource. Turning diagnostics on will allow you to configure some diagnostics settings, as shown in Figure 3-9.

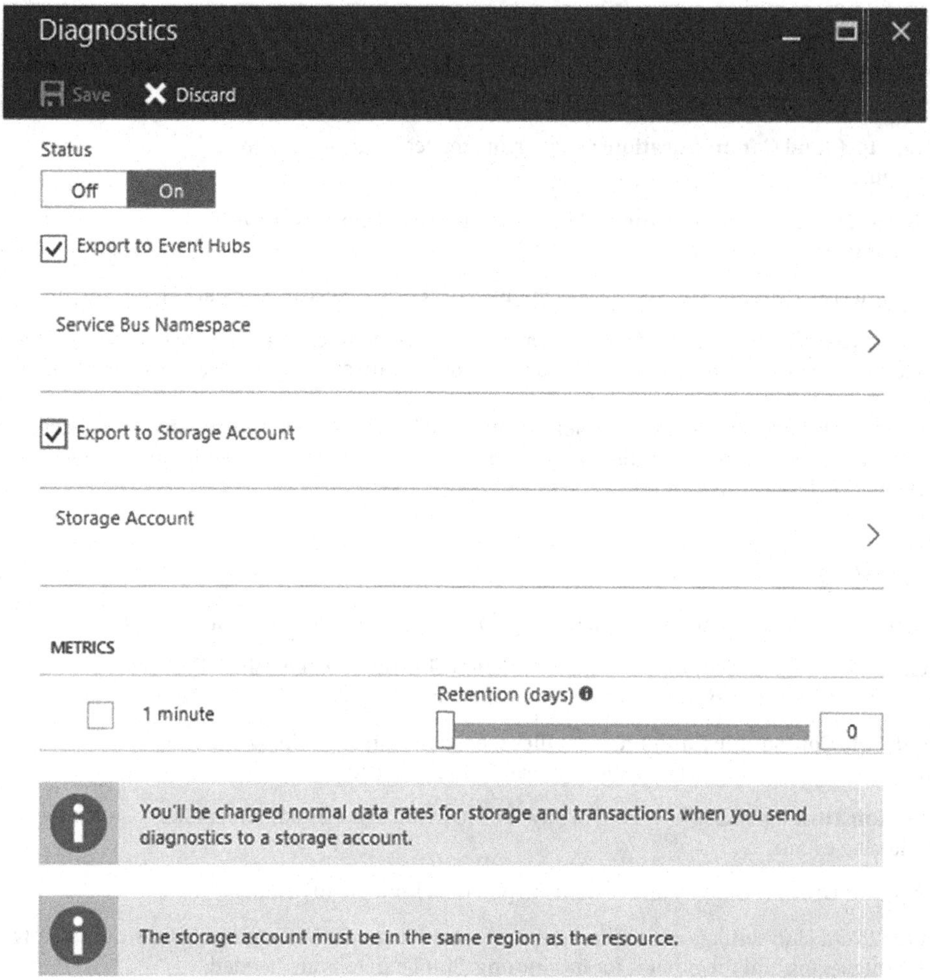

Figure 3-9. *IoT Hub diagnostic settings*

The diagnostics help provide better insight on the overall state of the resources in IoT Hub, such as the health of the services and the devices connected to your IoT Hub. From this information you can glean root-cause issues and proactively see what is happening within IoT Hub.

The first thing to do is to select what to do with the diagnostics data. Your options are to archive it to an Azure storage account, stream it to an Azure event hub, or send it to Log Analytics. From there, you'll need to specify the storage account, event, hub, or log analytics account. You can choose all three if you'd like. You are not limited to selecting a single option.

Other Settings

Wrapping up this section, there are a few more blades to cover.

Shared Access Policies

The Shared Access Policies blade allows you to define a set of permissions to endpoints. These policies can be applied to services, devices, and applications to define a fine-grained access control. When a device connects to IoT Hub, IoT Hub will authenticate the endpoints by verifying a token to both the defined shared access policies as well as the device identity registry security credentials.

New shared access policies are created by simply providing a name to the policy and then specifying the permissions: registry read, registry write, service connect, and device connect.

Pricing and Scale

The Pricing and Scale blade is similar and common to all services in Azure. It simply allows you to pick the performance level of Azure IoT Hub. Currently there are three tiers: Free, Standard S1, and Standard S2. The difference between then is simply how many messages you plan to send to the corresponding IoT Hub.

A couple of items to pay attention to: first, only one free tier can be created per IoT Hub. Second, IoT Hub does not let you transition between the free and standard tier. Meaning, if you have a free tier and want to upgrade it to a standard tier, this is not possible. The blade will inform you of that.

Summary

The goal of this chapter was to provide insight into the challenges of connecting devices to IoT solutions and how IoT Hub overcomes these challenges, including security and performance. This chapter provided an overview of Azure IoT Hub and then looked at why Azure IoT Hub is a viable and solid solution for solving the challenges and problems that IoT solutions face today.

Lastly, this chapter covered the creation and configuration of an Azure IoT Hub, including the different options and settings and what you need to know when creating and configuring your Azure IoT Hub.

However, there is more to do! So in the next chapter you will add your device to Azure IoT Hub and start sending messages.

CHAPTER 4

■ ■ ■

Ingesting Data with Azure IoT Hub

Chapter 2 walked you through generating data with the Raspberry Pi and the Tessel. Chapter 3 introduced you to the Azure IoT Hub, a fully managed communications service which provides secure, reliable, and scalable messaging between IoT devices and IoT solutions. Chapter 3 walked through the creation and configuration of an Azure IoT Hub, but no devices were registered. Chapter 2 created a solution but it didn't send any data to the IoT Hub. Thus, this chapter is going to plug those two together. Essentially, you'll first register your device with the IoT Hub and then modify your project to send data to Azure IoT Hub.

Registering the Device

Currently, the Azure portal does not provide the ability to provision or configure IoT Hub devices. However, as part of the Azure IoT SDK there is an application called Device Explorer which provides the means for managing devices and viewing device-to-cloud and cloud-to-device messages.

Thus, open your favorite browser and go to `https://github.com/Azure/azure-iot-sdk-csharp`. On the web page for the SDK, click the Download ZIP button. Once the zip file is downloaded, extract it to a folder. Within that folder, navigate to the `\tools\DeviceExplorer\` folder. The Device Explorer application is a Visual Studio solution, so open Visual Studio 2015 and open the DeviceExplorer solution.

Before running the solution, compile the solution first because it will need to download all the project dependencies from Nuget. Once the compile succeeds successfully, run the project.

In order to properly register a device in Device Explorer, you first need some information from the Azure portal. Log in to the Azure portal and open the IoT Hub that was created in the previous chapter. Notice in Figure 4-1 that in the Usage section of the Overview blade, there are no devices listed. But what you need from here is the connection string in order to connect the Device Explorer to IoT Hub in order to register any devices.

© Scott Klein 2017

S. Klein, *IoT Solutions in Microsoft's Azure IoT Suite*, DOI 10.1007/978-1-4842-2143-3_4

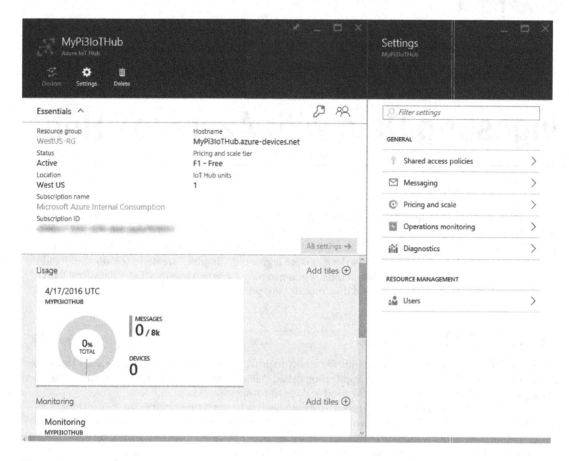

Figure 4-1. *IoT Hub overview*

In the Settings blade, click the "Shared access policies" option, and then in the Shared access policies blade, click the iothubowner policy. This will open the Access permissions blade for the iothubowner policy. In the Policy blade, the piece of information needed is the primary key connection string, highlighted in Figure 4-2. Click the Click to Copy icon to the right of the connection string field to copy the value to the clipboard.

The Device Explorer tool uses a set of Azure IoT service libraries to execute a number of tasks such as adding devices or viewing messages. It uses the connection string in order to connect to Azure IoT Hub via the service libraries (API calls).

The connection string is made of three pieces:

- **HostName:** The URI of the Azure IoT Hub, thus, MyPi3IoTHub.azure-devices.net

- **SharedAccessKeyName:** The policy name, in this case iothubowner

- **SharedAccessKey:** A unique key used to authenticate to Azure IoT Hub

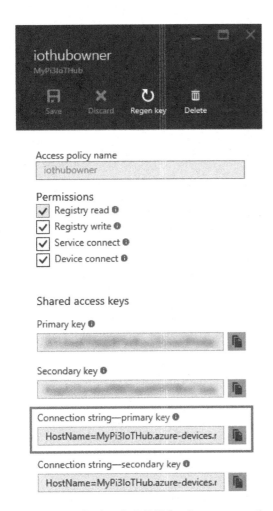

Figure 4-2. Getting the IoT Hub primary connection string

With Device Manager running, ensure that the Configuration tab is selected and paste the connection string into the "IoT Hub Connection String" text box. Click the Update button to save the configuration. You will then receive a message confirming the update of the setting and configuring the tool to communicate with IoT Hub. Device Manager should now look like Figure 4-3.

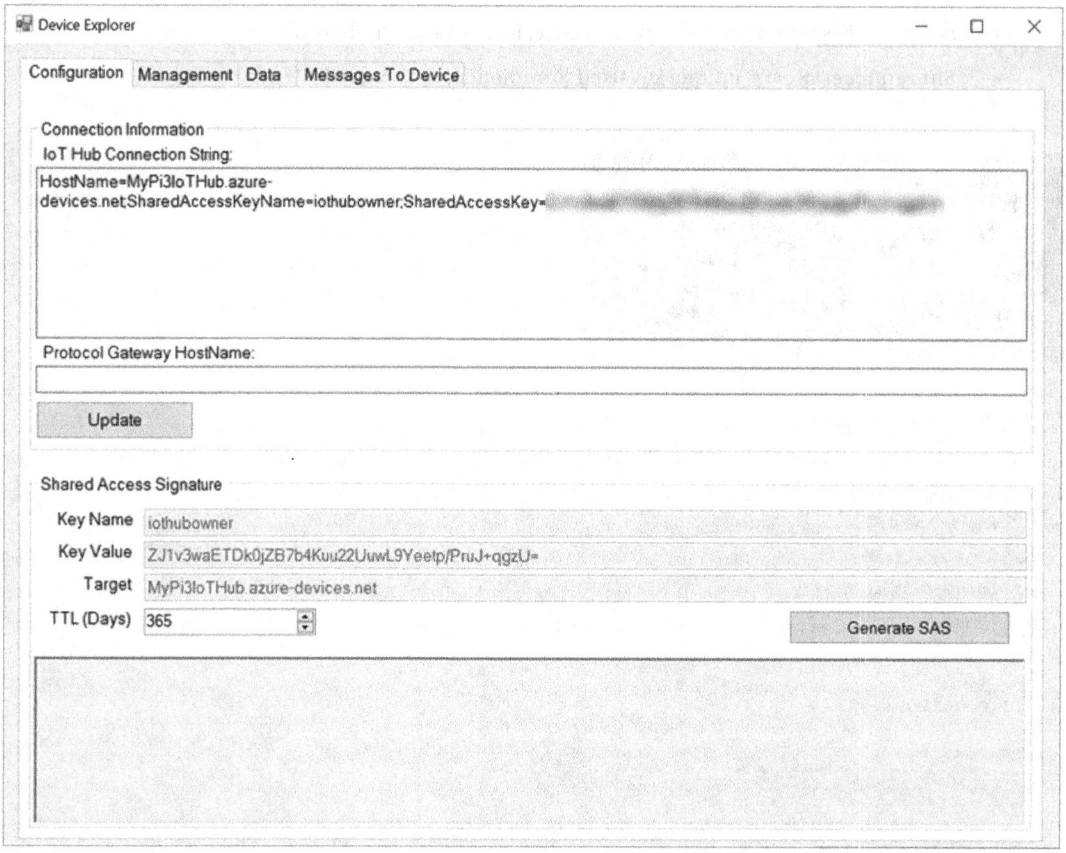

Figure 4-3. *Configuring the Device Explorer*

Next, click the Management tab. This tab is where devices are registered with IoT Hub. As shown in Figure 4-4, you can create, delete, and update devices with IoT Hub.

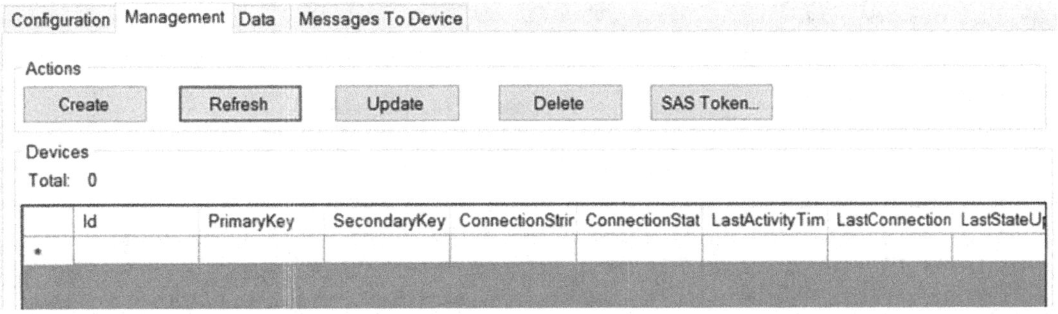

Figure 4-4. *Managing devices in the Device Explorer*

Click the Create button. This will open a dialog that will prompt you to provide a device name for the device you want to add, as seen in Figure 4-5. Associated with this device ID are two values, a primary key and a secondary key. These values will be prepopulated unless you specify to not have them autogenerated by unchecking the Auto Generate Keys checkbox.

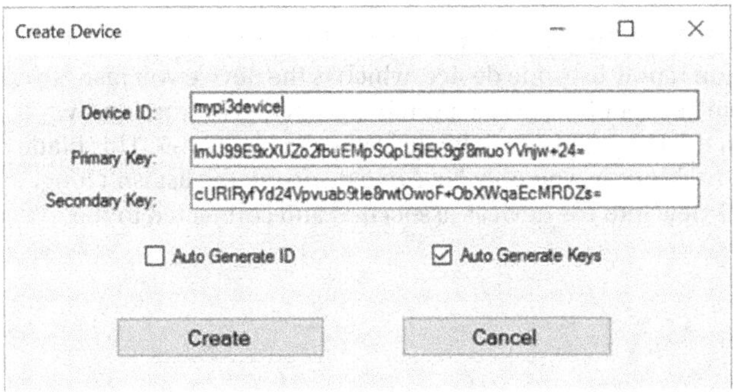

Figure 4-5. *Creating a new device*

Provide a device ID by entering a unique name and then click Create. For this example, I named my device "mypi3device." Obviously you will want to have a more unique and descriptive name, especially if you are dealing with more than one device.

Clicking Create on the dialog will then create the device and register it with Azure IoT Hub. Device Explorer will be updated with a list of all registered devices, as shown in Figure 4-6.

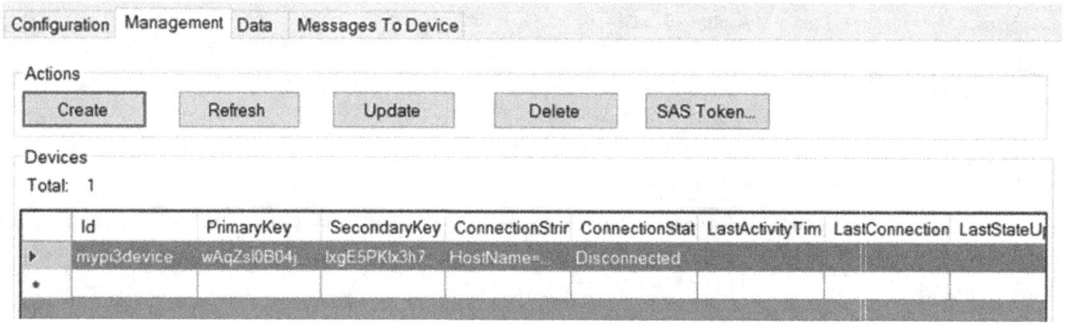

	Id	PrimaryKey	SecondaryKey	ConnectionStrir	ConnectionStat	LastActivityTim	LastConnection	LastStateU
▶	mypi3device	wAqZsl0B04j	lxgE5PKlx3h7.	HostName=...	Disconnected			
＊								

Figure 4-6. *Newly created device*

The information that should stand out in Figure 4-6 is the connection status, shown as Disconnected. Since your device isn't connected to the Azure IoT Hub, this status makes sense. You haven't added the code to your Visual Studio project to wire it up to the IoT Hub. You'll do that momentarily. However, in order to connect the app to the Azure IoT Hub you will need the connection string. This tool makes it easy: you simply right-click the row for the device and select "Copy connection string for selected device" from the context menu.

Before you modify the application, first go back to the portal and refresh the Overview blade. The first thing to point out is that in the usage section it now lists one device, which is the device you just added. The next thing to point out is that on the header of the Overview tab the Devices button is now enabled. Clicking the Devices button will open the Device Explorer blade, shown in Figure 4-7. This blade lists all the registered devices and their status. In this case, you only have one device so the list isn't long. However, this blade does provide a great initial view into the devices associated and connected to the corresponding IoT Hub.

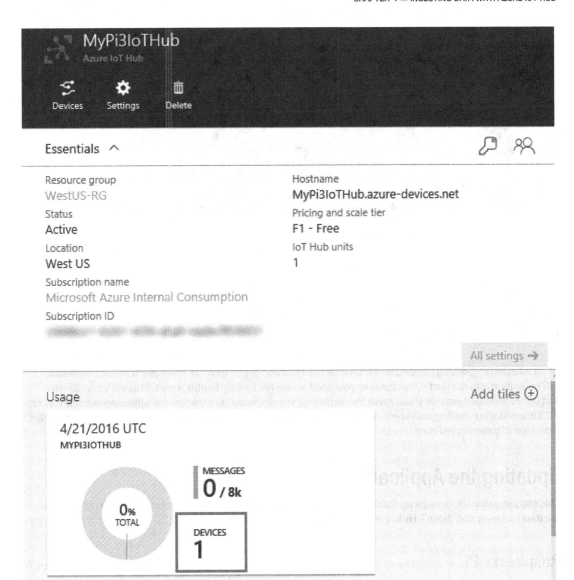

Figure 4-7. Updated Overview blade showing new device

There are several things that are nice about the Device Explorer blade. First, you can add additional columns to the list of devices. By default, it only shows the device name and status. However, you can add three additional columns: Status Reason, Last Status Update, and Last Activity. All three of these columns can provide real-time insight into the status of the device to help troubleshoot any connectivity issues the device might be having.

Second, it allows you to filter the list of devices, which will come in handy if you have many, many devices to manage. Clicking a specific device will open a Details blade, which provides key and connection string information as well as the ability to enable and disable the connection to the IoT Hub, as shown in Figure 4-8.

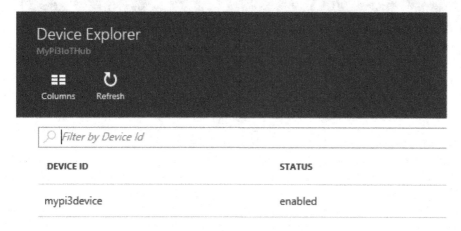

Figure 4-8. *Azure Portal Device Explorer*

Honestly, registering the device with Azure IoT Hub was pretty easy. At some point this functionality will be available in the portal, but for now you need to use the Device Explorer tool. This isn't necessarily bad because it does provide some great functionality. The next step is to update the application to connect to IoT Hub and start sending messages. You can then come back to the Device Explorer tool and refresh the list to show it in a connected state.

Updating the Application

This section will walk through updating both the Raspberry Pi and Tessel applications to add code to send the data to Azure and the IoT Hub. We'll begin with the Raspberry Pi.

Raspberry Pi

As it currently sits, the application created in Chapter 3 generates good data, but it does not send any data to Azure IoT Hub. The changes necessary to implement this functionality are actually quite minor and easy. The first thing to do is add the following line of code below the class declaration section with the other declared variables:

```
private const string IOTHUBCONNECTIONSTRING = "HostName=<IoTHubName>.azure-devices.net;DeviceId=<DeviceID>;SharedAccessKey=<SharedAccessKey>";
```

Next, replace the connection string with the connection string copied from the Device Explorer:

```
private const string IOTHUBCONNECTIONSTRING = "HostName=MyPi3IoTHub.azure-devices.net;DeviceId=mypi3device;SharedAccessKey=<SharedAccessKey>";
```

The connection string has three components:

- **HostName**: The name of the Azure IoTHub hostname. You can get this from the Azure portal for your IoT Hub Overview blade, shown in Figure 4-7.

- **DeviceID**: The name given to the device when registering the device with IoT Hub, shown in Figure 4-5.

- **SharedAccessKey**: The primary symmetric shared access key stored in base64 format. This is the same key found on the Device Details pane in the Primary Key field, discussed earlier.

The next step is to modify the code in the readSensor() method as follows:

```
DhtReading reading = await dht.GetReadingAsync().AsTask();
if (reading.IsValid)
{
    double temp = ConvertTemp.ConvertCelsiusToFahrenheit(reading.Temperature);

    var payload = new
    {
        Device = "Pi2",
        Sensor = "DHT11",
        Temp = temp.ToString(),
        Humidity = reading.Humidity.ToString(),
        Time = DateTime.Now
    };

    string json = JsonConvert.SerializeObject(payload);

    //send to listbox
    listBox.Items.Add(json.ToString());

    //send to azure IoT Hub
    Message eventMessage = new Message(Encoding.UTF8.GetBytes(json));
    DeviceClient deviceClient = DeviceClient.CreateFromConnectionString
    (IOTHUBCONNECTIONSTRING);
    await deviceClient.SendEventAsync(eventMessage);

}
```

I'll point out some of the changes in the code. First, you're still converting the temperature to Fahrenheit, but the next statement creates an object not only with the temperature and humidity, but you also want to track which specific device and sensor the temperature and humidity is coming from, as well as the date and time of the generated data.

You then use the JsonConvert.SerializeObject method to serialize the specific object to a JSON string. That JSON is then sent to the ListBox. Next, with the Microsoft.Azure.Devices.Client package installed from Nuget, you can add the code to send the temperature data to Azure IoT Hub. Thus, the three lines of code immediately following the ListBox statement do just that. The first line creates a new instance of the Message class, which the serialized JSON object is added to. The second statement creates an instance of the DeviceClient class in which a connection is made to Azure IoT Hub using the connection string added above. Lastly, the message is sent to Azure IoT Hub using the SendEventAsync method.

Now, good programming practice dictates that you create a class called DeviceMessage with the appropriate properties and then use a generic in Newtonsoft to serialize and deserialize the class. The code above is a quick solution to start sending data. However, to do it right, right-click the Temperature project and select Add ➤ Class. In the Add New Item dialog, ensure the Class template is selected and rename the class to DeviceMessage and click OK. Change the class DeviceMessage to public and then add the properties as follows:

```
public class DeviceMessage
{
    public string Device { get; set; }
    public string Sensor { get; set; }
    public string Temp { get; set; }
    public string Humidity { get; set; }
    public DateTime Time { get; set; }
};
```

Then, back in the readSensor() method on the MainPage, replace the var payload and string json statements with the following:

```
DeviceMessage msg = new DeviceMessage
{
    Device = "Pi2",
    Sensor = "DHT11",
    Temp = temp.ToString(),
    Humidity = reading.Humidity.ToString(),
    Time = DateTime.Now
};
string json = JsonConvert.SerializeObject(msg, Formatting.Indented);
```

So now you are following good coding practice (well, technically, you would also add some error handling). At this point, the solution is ready to be compiled and run. Recompile the solution and when that is done, deploy the solution to the Raspberry Pi by clicking the green Play button next to Remote Machine on the toolbar again. Or, simply press F5.

If the ListBox line of code is uncommented, you'll see data written to the ListBox. However, the important part is seeing it in Azure IoT Hub. Thus, go back to the Device Manager tool and select the Data tab. On the Data tab, click the Monitor button. If the IoT Hub is indeed receiving messages, you'll see them showing up in the Event Hub Data ListBox, as shown in Figure 4-9. The Cancel button stops messages from being written to the ListBox; it does not stop messages from being received by IoT Hub. The Clear button clears the ListBox.

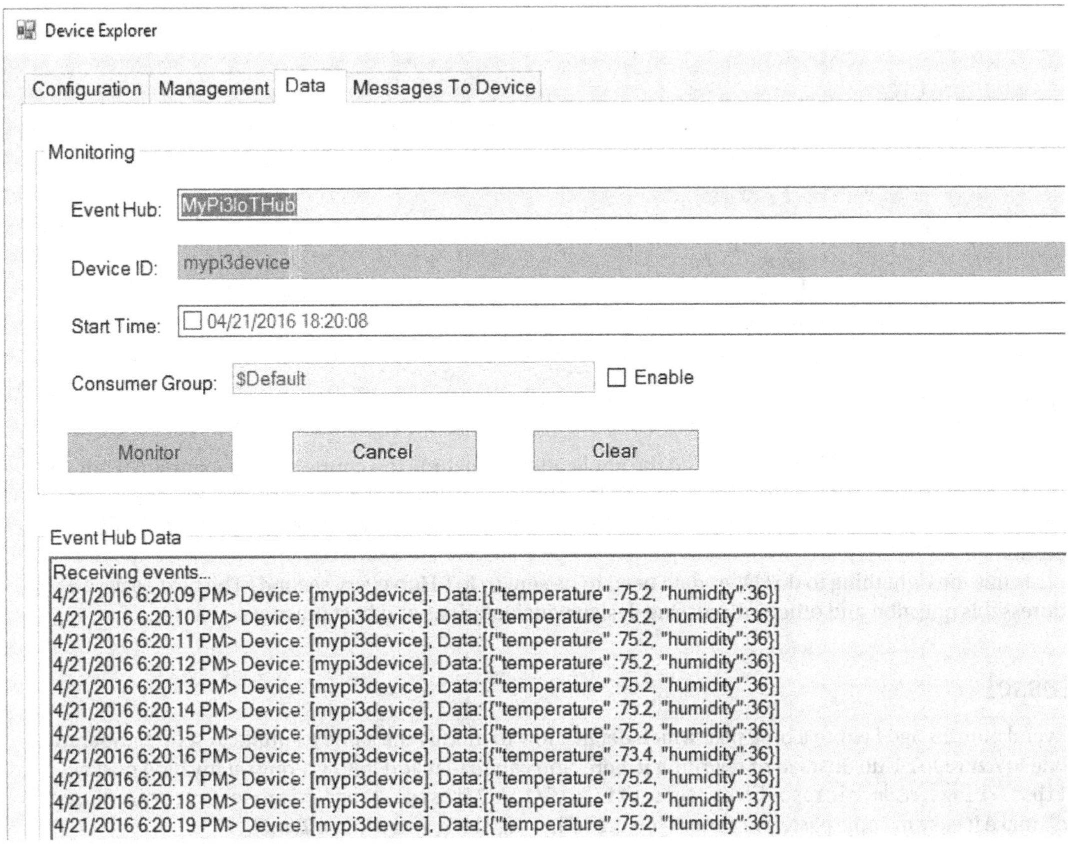

Figure 4-9. *Viewing live data in Device Explorer*

Selecting the Management tab and clicking the Refresh button now shows the ConnectionState as Connected, which is seen in Figure 4-10.

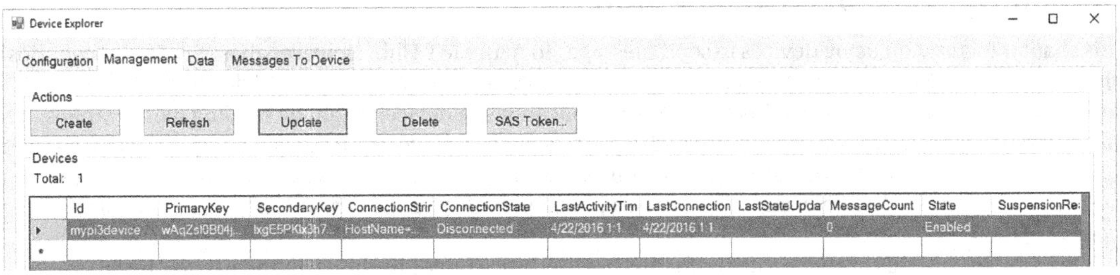

Figure 4-10. *Device Explorer connection state*

Now go back to the Azure portal and refresh the Overview blade; you should now see in the Usage section the number of messages ingested into Azure IoT Hub at the time of the refresh. The blade will refresh automatically every minute so you will get an idea of how many messages are being sent and consumed by IoT Hub (see Figure 4-11).

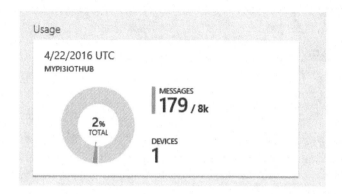

Figure 4-11. Updated message count in the Azure Portal

In this example, you simply modified the application to include the connection to Azure IoT Hub and sent the temperature and humidity messages to IoT Hub. The temperature and humidity readings were taken every second, and immediately sent to IoT Hub, which means data was being sent to IoT Hub every second.

Is this the right thing to do? Does data need to be sent to IoT Hub every second? The next section will address this question and others, discussing design considerations and best practices.

Tessel

Several months ago I wrote a blog post which details how to modify the same example to send temperature code to Azure IoT Hub. Instead of rewriting it here, you can read it and follow along on my blog post at `https://blogs.msdn.microsoft.com/sqlscott/2016/10/17/tessel-2-and-microsoft-azure/`. Plus, it will make it easy to copy/paste the code if you are following along and using a Tessel.

Considerations

Before wrapping up this chapter I'd like to discuss a couple of additional items that are relevant to this topic.

Uploading Files to Azure IoT Hub

This chapter focused on using devices to send messages to Azure IoT Hub. However, there are some scenarios where it is more difficult to map the data from the device into the comparatively small device-to-cloud messages received by IoT Hub. For instance, some applications or devices might be sending files or videos that would typically be processed in batches with services such as Hadoop/HDInsight or Azure Data Factory.

One example is `how-old.net`, a site that captures your picture and uses Azure Machine Learning to guess how old you are. A simple, and silly, example, but it provides a good example of getting real-time analysis of data.

In cases such as this example, files uploaded from a device or application can still use IoT Hub, especially to take advantage of the security and reliability functionality, meaning files are stored in Azure Blob Storage but they are managed via Azure IoT Hub, thus receiving the benefits provided, including notifications of uploaded files.

As shown in Figure 4-12, configuring Azure IoT Hub to work with uploaded files from devices is quite simple. In the Properties page of IoT Hub, select the File Upload link and then simply select the Azure Storage account and container in which the blobs will be uploaded, and then select whether to receive notifications of uploaded blobs and notification settings.

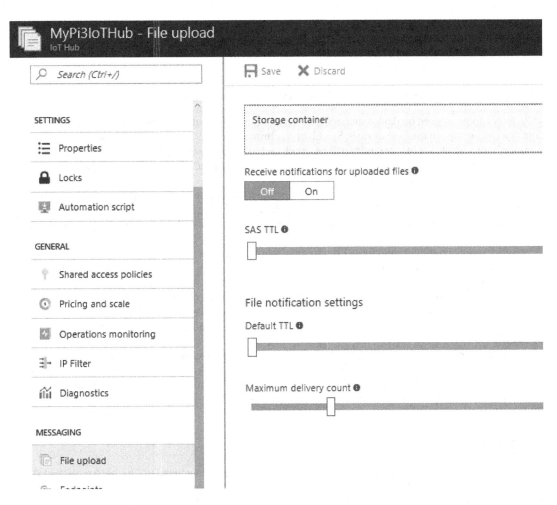

Figure 4-12. *Configuring IoT Hub file uploads*

The code for uploading files is quite simple. The following method specifies a file, creates a connection to Azure IoT Hub, and then, like the previous example, uses the UploadToBlobAsync method on the DeviceClient class to upload the file to Azure Blob Storage:

```
private static async void SendToBlobAsync()
{
    string fileName = "<filename>";

    DeviceClient deviceClient = DeviceClient.CreateFromConnectionString
    (IOTHUBCONNECTIONSTRING);

    using (var sourceData = new FileStream(@"<filename>", FileMode.Open))
    {
        await deviceClient.UploadToBlobAsync(fileName, sourceData);
    }
}
```

Similarly, with a few lines of code your application can receive file upload notification messages from Azure IoT Hub. Both of these examples illustrate the ability and flexibility of device-to-cloud and cloud-to-device messaging functionality of IoT Hub.

Other Device Management Solutions

While this chapter focused on device management using Microsoft solutions, there are many open source tools that can be used to manage devices. Search in any browser search engine for "iot device management solutions" and a whole list of companies will be listed, including Opensensors.io, Allegro, Bosch, and Kaa. In fact, this list is just a small sampling. In a recent report, nearly 20 companies provide IoT device management, with Microsoft being one of them.

What makes many of these solutions interesting is that they are open source and allow you to deploy them yourself to virtual machines. With Kaa, for example, you can create a sandbox locally for development and experimentation on a small scale, and the sandbox can be either installed locally or deployed to a virtual machine.

A quick search on the Internet for IoT device management solutions comparison will turn up a few reports to look at, comparing security, protocols, and integration features. Some of the reports require you to pay to view them, but there are a few that are free and they do a decent job in their comparisons.

Most the solutions have HTTP/S and MQTT data collection protocol support, while only a few support AMQP. The majority of them offer REST API integration and also provide some flavor of real-time analytics. The support for real-time analytics is provided by a wide range of solutions including Apache Storm or a rules engine.

While many of these device management solutions will provide security, data collection protocols, and some analytics, the key point is that the Azure IoT Hub tracks events across categories of operations, allows you to monitor the status of operations within the hub, and integrates very well with Azure Stream Analytics for real-time streaming data insight, which, by the way, is the topic of the next chapter.

Summary

This pivotal chapter focused primarily on connecting the application developed in Chapter 2 with the IoT Hub created in Chapter 3. While the application was a simplistic one, you saw how easy it was to make the minor changes needed to the application in order to send messages to Azure IoT Hub.

In the real world, it will probably take a bit more work, but the goal with these chapters was simply to illustrate the process for connecting your IoT device to an IoT solution.

Lastly, this chapter discussed some design considerations and best practices for device-to-cloud communication, including device pull vs. send frequency and others.

■ ■ ■

CHAPTER 5

Azure Stream Analytics

At this point in the process, the data from the devices is sitting in Azure IoT Hub waiting for the next step of the journey. In this case, "sitting" means that the data has just arrived in Azure IoT Hub and is waiting to be picked up by another service for processing. Chapter 3 walked through the process of creating and configuring an Azure IoT Hub to receive the messages sent from the devices. And, depending on how the IoT Hub was configured, the data currently sitting in IoT Hub could have a very short lifespan.

The configuration of the IoT Hub in Chapter 3 chose the free tier, which means that the retention time for the messages is only a day. A single day. No more. Picking a standard S1 or standard S2 tier gives the flexibility to up the retention time to a maximum of seven days. A whole week. Honestly, whether it's a single day or seven days, that is not a bad thing. The data shouldn't be sitting around at this point anyway. The fact that IoT Hub keeps the data around for up to seven days means that it is guaranteeing reliable processing while also helping to absorb peak loads.

However, this point goes back to the *hot path/cold path* discussion. Really, you should have already determined what you intend to do with the data. Do you need real-time or near real-time analysis of the data, or do you plan on storing the data and keeping the data around for future analysis?

The thing is, regardless of the answer, IoT Hub is only a temporary holding place for incoming data. The next step is to pick up that data as soon as it comes in and do something with it, whether it's immediate analysis or storing it long term. In addition, regardless of the path (hot or cold), the data needs to be picked up and processed, so what is needed is a real-time processing engine that can pick up the data from IoT Hub, drop it where you want it, and in between maybe do some data transformations. And it would be nice if this engine was also highly scalable.

Luckily there is such an engine, and it is called Azure Stream Analytics. This chapter will take a detailed look at this service, what it can be used for and the scenarios in which it applies, and how to create and configure a Stream Analytics service.

What Is Azure Stream Analytics?

Simply put, Azure Stream Analytics (ASA) is a real-time event processing engine that provides analytical capabilities on streaming data from devices, applications, sensors, and more. It is a fully managed, highly scalable Azure service focused on making access to deep data insights powerful, inexpensive, and easy to use.

In the big picture, Azure Stream Analytics picks up ingested data from one of several sources, does some level of data transformation, and outputs the result to a different set of destinations. The flow that this book is taking is shown in Figure 5-1, which, as you saw in the previous chapters, is sending data from the devices directly to Azure IoT Hub. As discussed previously, Azure provides the option to use Cloud or Field gateways into IoT Hub for legacy (custom protocols) or low-powered devices, enabling them to take advantage of the same workflow and analysis benefits.

© Scott Klein 2017

S. Klein, *IoT Solutions in Microsoft's Azure IoT Suite*, DOI 10.1007/978-1-4842-2143-3_5

Later chapters will fill in and complete the picture, but for now, Figure 5-1 shows where Stream Analytics fits into the current process.

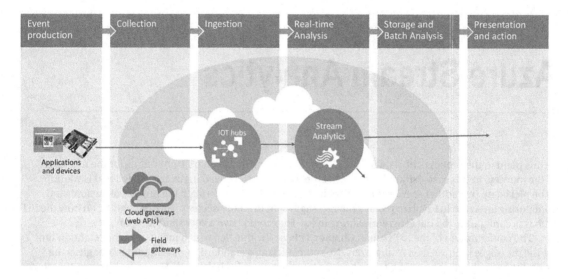

Figure 5-1. *Data flow*

Given what you know up to this point, think about the amount of data that can flow into Azure IoT Hub and get picked up by Azure Stream Analytics to be processed and acted upon in real time. Imagine the different scenarios where this can be useful: anomaly or fraud detection, real-time stock trading, or even high-stakes betting and gambling. What all of these scenarios have in common is the need to be scalable, reliable, and cost effective in every aspect of the process.

Azure Stream Analytics isn't just limited to the scenarios just mentioned. Stream Analytics fits anywhere businesses need the ability to gain real-time insights into incoming data. At the beginning of this chapter, the topic of hot path and cold path data processing was mentioned. Hot path equates to real-time processing whereas cold path equates more to micro-batch processing. Real-time processing denotes the processing of an infinite stream of input data, with the time between data ingestion and ready results being small, typically measured in seconds.

Micro-batch processing is the method by which a group of transactions are collected over a period of time and processed as a batch. In Azure Stream Analytics, micro-batch processing is performed by a Stream Analytics job in which an input and output are defined and a SQL query is used to process the incoming data. An example of this will be shown in the next chapter as you continue the example from Chapter 4.

With the data in Azure IoT Hub, the next step in the journey is to create and configure a Stream Analytics job. However, before I get to that, let's take a step back and drill down into the key capabilities and benefits of Stream Analytics, which will help put into perspective the scenarios where Stream Analytics fits as well as the best way to configure the Stream Analytics job.

Key Benefits and Capabilities

The capabilities and benefits of Azure Stream Analytics share many of the character traits as Azure IoT Hub and other services which have been discussed and which will be discussed in chapters to come.

- **Scale**: With hundreds and thousands of devices sending vast amounts of data into the cloud, Azure Stream Analytics is designed and built to handle the millions of events per second coming in, up to 1GB/second. What makes this possible is the partitioning capabilities of Event Hubs.

- **Reliability**: Data loss prevention and business continuity, two key facets of Stream Analytics, are provided via built-in recovery features and the ability to maintain state. As such, Stream Analytics is able to archive events and reapply processing to ensure repeatable results.

- **Connectivity**: Stream Analytics allows connectivity to many different sources and destinations, including Event Hubs and IoT Hubs for stream ingestion, as well as Azure Storage (Tables and Blobs), Azure SQL Database, Azure Data Lake, and even Power BI for output and results. A full list will be discussed in Chapter 6.

- **Ease of Use**: Not only is creating inputs and outputs extremely simple, but Stream Analytics also makes data transformations simple through the support of a variant of the SQL language called the Stream Analytics Query Language. Available right in the Azure portal, this language supports IntelliSense and auto-complete directly in the query editor.

- **Cost Effectiveness**: As with other Azure cloud services, Stream Analytics is designed to provide a high-value, real-time analytics solution for a low cost. Built around a pay-as-you-go model, the cost is based on the number of units and the amount of data processed.

The bullets above summarize the value of Stream Analytics, but the proof of its value is clearly seen once it is put into action. As you'll see in the next chapter, many of the capabilities and characteristics of Stream Analytics will stand out as they are put to use.

One such case is the query language. The SQL-like query language includes many of the familiar functions, filters, and other operators found in SQL Server. Another example is the ease of creating inputs and outputs simply with a few clicks in the portal. But let's be clear that the portal isn't the only place jobs can be created. Azure Stream Analytics has a .NET API that enables entire job management, such as creating, configuration, and running complete Stream Analytics jobs. There are also nearly 20 Azure PowerShell cmdlets to automate common Azure Stream Analytics tasks, such as starting and stopping jobs, creating job input, and more.

That is enough background and information to get started. It is time to create and configure your Stream Analytics job and put it to use. The following section will walk through the creation of a Stream Analytic job and discuss the different configurations and settings available.

Creating an Azure Stream Analytics Job

To begin, open your favorite browser and go to Azure.Portal.com and sign in. Once in the portal, click New, select Internet of Things, and then select Stream Analytics Job, as shown in Figure 5-2.

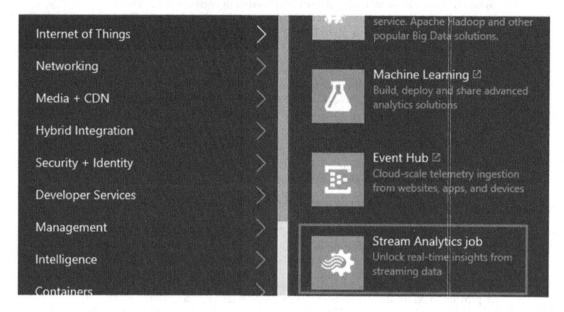

Figure 5-2. *Creating a new Stream Analytics job*

If you are familiar with the previous version of the portal, it was just called Stream Analytics, but in the new portal (**portal.azure.com**) the name is Stream Analytics job.

The New Stream Analytics Job blade will appear, similar to the one in Figure 5-3. In the blade, enter or select the following:

- **Job Name**: The name of the Stream Analytics job. In my demo, I named it MyPiStreamAnalytics.

- **Resource Group**: Either select an existing Resource Group or create a new one. Be sure to select the same resource group that your IoT Hub is in.

- **Location**: Select the appropriate region in which to host the IoT Hub.

One thing to notice that is not on this blade is the pricing and scale option to initially set the performance and cost. Once the Stream Analytics job is created, the performance level will be configurable.

Click the "Pin to dashboard" checkbox and then click Create. The creation of the Steam Analytics job will take only a moment, probably less than a minute.

Figure 5-3. *Configuration of the new Stream Analytics job*

Once it is created, you will see a new icon on your dashboard. Click the new Stream Analytics job icon on your dashboard to open up both the Overview blade and the Settings blade for the newly created Stream Analytics job, similar to Figure 5-4.

The left Overview blade provides a summary of the Stream Analytics job as well as the all-important job topology information such as the number of inputs and outputs, queries, and vital job monitoring information.

Two important items to point out on the Overview blade are the Start and Stop buttons for the job. When the job is first created, those buttons are disabled simply because no job is configured, so no inputs or outputs are defined and no query is created. Therefore, the job cannot be run, so the buttons are disabled. Once an input, output, and a query have been defined, those buttons will be enabled.

There is also a Functions button that is disabled upon job creation. Functions allow you to call a REST endpoint, which allows each message received to be processed by a web service GET/POST. It allows the "self-discovery" of the endpoint parameters. Functions are primarily intended for use with AzureML (you will see this in the Azure Machine Learning chapter) where machine learning models can be enabled and invoked through a REST API.

Before getting to the Settings blade and the different configurations, a few minutes needs to be spent on monitoring and diagnostics. In Figure 5-4, there is no settings option for diagnostics. To turn on monitoring and diagnostics, click anywhere in the Monitoring graph, which will open the Diagnostics blade. In this blade is where you simply turn on or off diagnostics monitoring. When turning it on, it will ask you to pick the storage in which to store the telemetry and diagnostics data. Simply pick a storage account, ensuring that the storage selected is in the same data center as the SA job, and click Save.

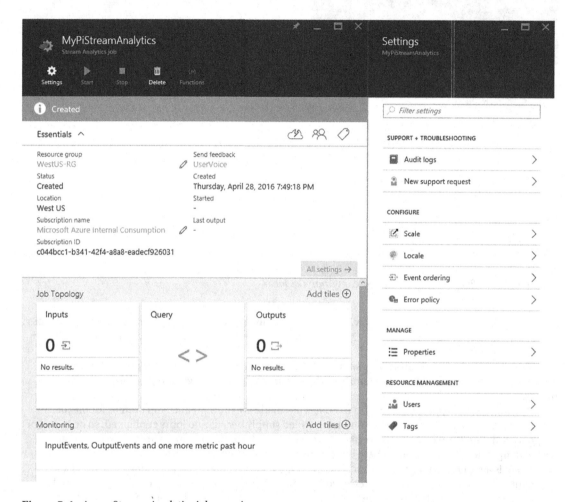

Figure 5-4. *Azure Stream Analytics job overview pane*

At this point, monitoring is turned on but no diagnostic telemetry data is being gathered because no metrics have been defined. To define metrics and alerts and begin gathering diagnostic data, click again in the Monitoring graph, which will open the Metric blade, shown in Figure 5-5.

At this point, there is no data because no inputs or outputs have been defined. Also, no alerts have been added, thus the "No available data" message in the Metric graph.

InputEvents, OutputEvents and one more metric past hour

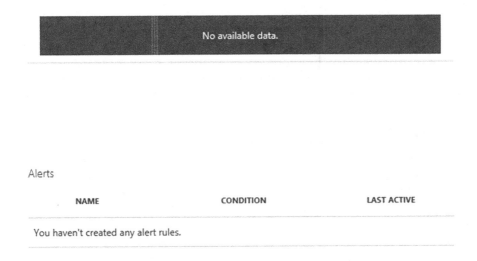

Alerts

NAME	CONDITION	LAST ACTIVE

You haven't created any alert rules.

Figure 5-5. *Azure Stream Analytics Metric pane*

However, it is worth at least mentioning here how to create and configure alerts so that when the inputs and outputs are created in the next chapter, it will be easy to come back here to configure the monitoring.

Besides providing a name, an alert is created by specifying which metric to monitor and the conditions for that metric. As shown in Figure 5-6, there are 11 key performance metrics that can be used to monitor and troubleshoot job and query performance along with their corresponding condition, threshold, and the amount of time over which to monitor the metric data.

Additionally, alerts can be raised via email to Azure subscription owners and contributors along with additional persons specified.

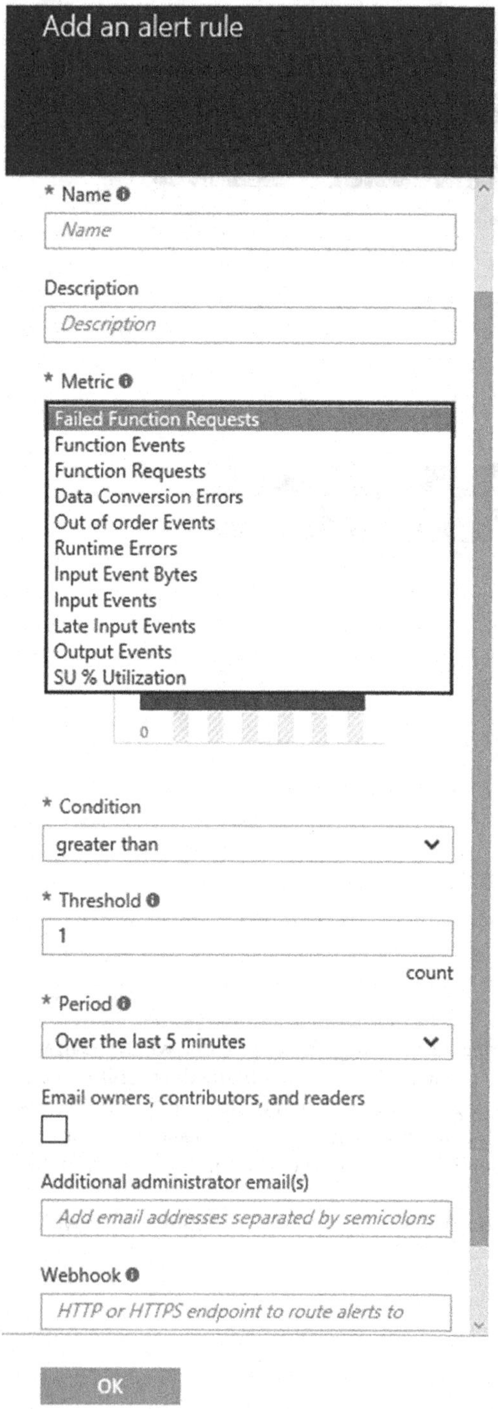

Add an alert rule

* Name ❶

Name

Description

Description

* Metric ❶

Failed Function Requests
Function Events
Function Requests
Data Conversion Errors
Out of order Events
Runtime Errors
Input Event Bytes
Input Events
Late Input Events
Output Events
SU % Utilization

0

* Condition

greater than ⌄

* Threshold ❶

1

count

* Period ❶

Over the last 5 minutes ⌄

Email owners, contributors, and readers

☐

Additional administrator email(s)

Add email addresses separated by semicolons

Webhook ❶

HTTP or HTTPS endpoint to route alerts to

OK

Figure 5-6. *Defining Azure Stream Analytics alerts*

The current list of available metrics helps gain insight into what is happening within the Stream Analytics job through error and performance tracking. The following lists the available metrics and their definition:

- **Failed Function Requests**: The number of failed HTTP response codes (e.g. HTTP 400/500) from a function request.

- **Function Events**: The number of events that are triggered from a function request.

- **Function Requests**: The number of requests to each and all registered functions.

- **Data Conversion Errors**: The number of data conversion errors incurred by a SA job.

- **Out of Order Events**: The number of events that were received out of order, which were either dropped or given an adjusted timestamp.

- **Runtime Errors**: The number of errors incurred during a SA job execution.

- **Input Event Bytes**: Amount of throughput data received by the job in terms of bytes.

- **Input Events**: Amount of data received by the job.

- **Late Input Events**: The number of events arriving late from the source, which have either been dropped or have an adjusted timestamp.

- **Output Events**: Amount of data sent by the SA job to the output target.

- **SU% Utilization**: The utilization of the streaming units assigned to a job from the Scale tab.

Each metric has a corresponding condition and threshold. For example, you might select the Input Events metric with a condition of "greater than or equal to" and a threshold of 10,000 to closely monitor how much data your input is receiving. You might consider combining this metric with the SU% Utilization metric to ensure that you are getting optimal performance and not peaking the limits. These metrics together can tell you if you need to increase your streaming units to handle the number of events without pegging your SU% utilization.

You can have only six metrics displayed on the chart, so be wise with the metrics you choose to monitor. If your Stream Analytics job is pulling in data from multiple streams (from multiple IoT Hubs) for example, you might consider adding the Out-of-Order Events metric to determine if there is any latency of arriving messages. More on Out-of-Order Events shortly.

Scale

One of the most difficult aspects of configuring an Azure service is figuring out how many resources are needed for optimum performance. In Azure Stream Analytics, performance is determined by selecting the appropriate streaming units. In the Settings blade, click the Scale option, which opens the Scale blade and the option to select the number of streaming units, shown in Figure 5-7.

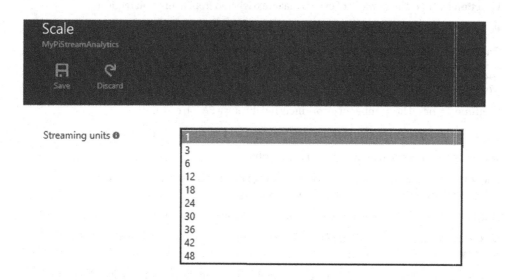

Figure 5-7. *Configuring Azure Stream Analytics streaming units*

Streaming units characterize the resources and performance of executing a Stream Analytics job. They embody a blended combination of memory, CPU, and I/O. Ultimately, each SU equates to roughly 1MB/second of throughput.

The trick is determining how many SUs are needed for a particular job. It's not rocket science, but it does take some knowhow along with some testing and monitoring. The correct SU selection will depend on the query defined for the job as well as the partition configuration for the inputs. Thus, when writing the query, how it is written will have an impact on the overall performance.

Another way, and possibly an easier way, to determine SUs is to test the lowest common denominator as a scale unit and understand the data volume peaks. IoT Hub will buffer the messages so you can have a lower value of SU and a higher retention period, which would equate to a process that is just slightly less than real-time but with a lower cost.

In essence, you are trying to find the balance of price vs. performance. Can you pay less for "not quite" real time and be fine with a 5 second processing window, for example? Or, do you need the SUs to provide the absolute real-time processing with high performance? The point is that getting the performance out of Azure Stream Analytics (i.e. increasing stream data processing) depends on the partition count for the inputs and the query defined for the job. You will probably spend some time tweaking both to find the right configuration. However, as stated earlier, test with the lowest common denominator and tweak from there based on your data volume peaks.

Event Ordering

Earlier it was stated that Stream Analytics was built to handle the millions of events per second coming in, up to 1GB/second. To achieve such numbers would mean that there are many devices sending a large amount data for Stream Analytics to process. This could also mean that quite possibly Stream Analytics is processing streams of data from multiple IoT Hubs or Event Hubs. SU is also based on volume of data processing and as such Stream Analytics scales nicely because device payloads are typically small, which is why it can process millions of messages on tumbling windows.

The complexity of these types of solutions might mean that some events don't make it from the client to the hub as fast as they should due to varying latency reasons, thus arriving at the hub out of order. Events might even arrive out of order once they've made the trip from the input source to Stream Analytics, especially in a multi-stream scenario. Additionally, to handle the multi-stream scenario, the complexity of the query increases to handle the arrival of late, independent events.

It is worth pointing out that AMQP defines a client side "buffer," which should preserve the order based on connectivity loss. This will even out the latency and make this more batch-enabled.

As such, consideration of what to do with these events must take place, which is the purpose of the Event Ordering blade, shown in Figure 5-8.

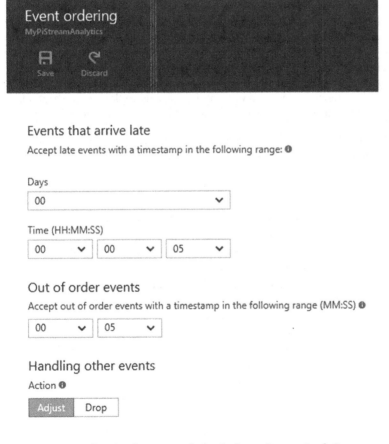

Figure 5-8. *Configuring the event ordering in Azure Stream Analytics*

The first section in this blade allows you to specify which delayed or late events to accept within a given timeframe. These events can still be picked up by Stream Analytics as long as a timespan is specified in days or hours/minutes/seconds. A delay in the case of late events is defined as the difference between the event's timestamp and the system clock.

The second section deals with events that are out of order during their trip from the input source to Stream Analytics. A likely scenario would be when you are combining multiple streams. Here, the choice is to accept the events as is, or to pause for a set period of time to reorder them.

The last section deals with events that arrive outside of the times specified in the previous sections. Here the choices are to simply delete the events or to adjust the events by changing their timestamp.

Event ordering is dependent on the record timestamp, and one of two timestamps are applied and used. By default, events are timestamped based on their arrival time to the input source. For Event Hub and IoT Hub, the timestamp is the arrival time when the event was received by the hub.

However, some applications necessitate the exact timestamp of when the event occurred. For these cases, the TIMESTAMP BY clause can be added to the SQL query, allowing for the use of custom timestamp values. For these cases, the timestamp value can be any field from the event data payload.

When TIMESTAMP BY is not included in the query, the timestamp from IoT Hub or Event Hub is used.

Audit Log

The Audit Log blade provides insight into what is happening with your Steam Analytics job. This information can be valuable when debugging jobs with facts regarding job status, job failures, and the status of currently executing jobs.

Audit information isn't just available to Stream Analytics. All Azure services provide detailed logging information for debugging and management. The Audit blade shows events from the past seven days, as shown in Figure 5-9. By default, a filter is applied, which shows events for the subscription, resource group, resource type (in this case, streaming jobs), given resource (in this case, Azure Stream Analytics jobs), and all events.

To change the filter, simply click the Filter button on the toolbar. While you can't change the severity, you can change the level of logging and the timespan in which events are tracked. By default, four types of events are tracked:

- Critical

- Error

- Warning

- Informational

As stated above, events are tracked for seven days by default. You can change this by selecting the Time span option in the Filter blade and selecting one of the following:

- Past 1 hour

- Past 24 hours

- Past week

- Custom

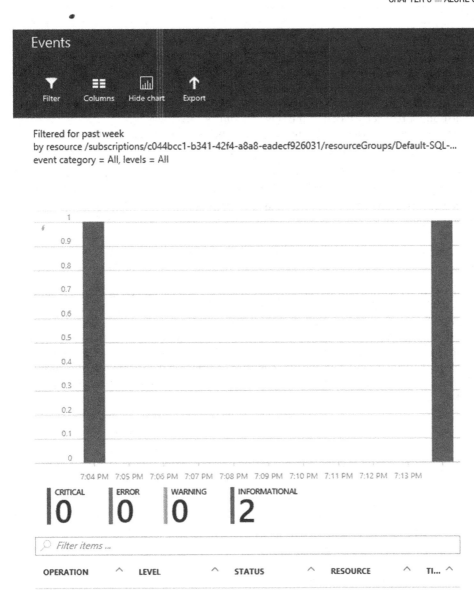

Figure 5-9. Azure Stream Analytics Events blade

The Events blade also lists log entries toward the bottom of the blade. You can see details of the log entry by clicking the specific event, which will open the Detail blade for that event. The Detail blade simply provides information regarding the specific event, such as event status, event level, the timestamp of the event, and more.

Additional Settings

The other configuration settings on the Settings blade allow you to initiate a support request, change the internationalization preference, view the Stream Analytics properties, define additional users, and configure how events are handled that fail to be written to the output.

Of all of the settings just mentioned, the Locale blade is the most important because it tells Stream Analytics how to parse and sort the data based on the language and locale settings.

The Error Policy is another one that shouldn't be overlooked, but isn't too critical. The Error Policy simply provides a Drop or Retry setting for failed events. The Drop option drops the event that caused the error completely while Retry retries the event until is succeeds.

Lastly, the Support Request blade will come in handy will working with critical errors found in the audit logs. Simply taking the information found in the audit logs for critical and error information and passing it along in the support request will provide a fast turnaround time for troubleshooting Stream Analytic issues within your Azure subscription.

Summary

This chapter provided an overview of Azure Stream Analytics, beginning with a discussion about streaming data and its lifespan, followed by a look at Azure Stream Analytics, what it is, its core capabilities and key benefits, and why it is considered a valid cloud streaming solution for real-time analysis of streaming data.

The last part of the chapter created a Stream Analytics job and then looked at the crucial configuration settings and some considerations when first setting up your Stream Analytics job. The following chapter will continue the example from Chapter 3 by taking the next configuration steps in the Stream Analytics process by authoring a job by specifying an input, output, and query settings to move the data currently sitting in Azure IoT Hub along.

CHAPTER 6

■ ■ ■

Real-Time Data Streaming

Chapter 5 introduced Azure Stream Analytics and its great capabilities and key benefits for real-time data streaming. That chapter also walked through creating a Stream Analytics job and looked at many of the settings for configuring a Stream Analytics job. It is time to move the data along to the next part of its journey and implement what was learned in Chapter 5.

As a recap, data is being generated by several devices and sent into Azure IoT Hub. This chapter will apply what was learned in Chapter 5 and connect Azure IoT Hub and Stream Analytics, allowing the data to continue its journey. To do that, you'll configure the Stream Analytics job created in Chapter 5 to create an input and output source, and then define the query which moves the data from Azure IoT Hub to a selected destination.

Configuring the Stream Analytics Job

With the data sitting in Azure IoT Hub, the next step in the process is to use Azure Stream Analytics to pick the data up from Azure IoT Hub and move it to where is needed for further processing. In this case, you'll pick the data up from Azure IoT Hub and drop it in Azure Data Lake Store. Azure Data Lake Store will be discussed in detail in the next chapter, but in this chapter you'll quickly create one so the data has a destination.

If the Azure portal isn't open, go to your favorite browser and navigate to `portal.azure.com` and log in. Once the portal loads, the MyPiStreamAnalytics tile for the Stream Analytics job created in the last chapter should be on the dashboard if you selected the "Pin to dashboard" option when creating the Stream Analytics job.

If "Pin to dashboard" was not selected, you can find the Stream Analytics job by searching for it at the top of the portal. In the Search Resources search box, type *stream analytics*, and select the Stream Analytics jobs option that will appear in drop-down list of search options. All existing Stream Analytics jobs, including the MyPiStreamAnalytics job, will be listed, as shown in Figure 6-1.

© Scott Klein 2017

S. Klein, *IoT Solutions in Microsoft's Azure IoT Suite*, DOI 10.1007/978-1-4842-2143-3_6

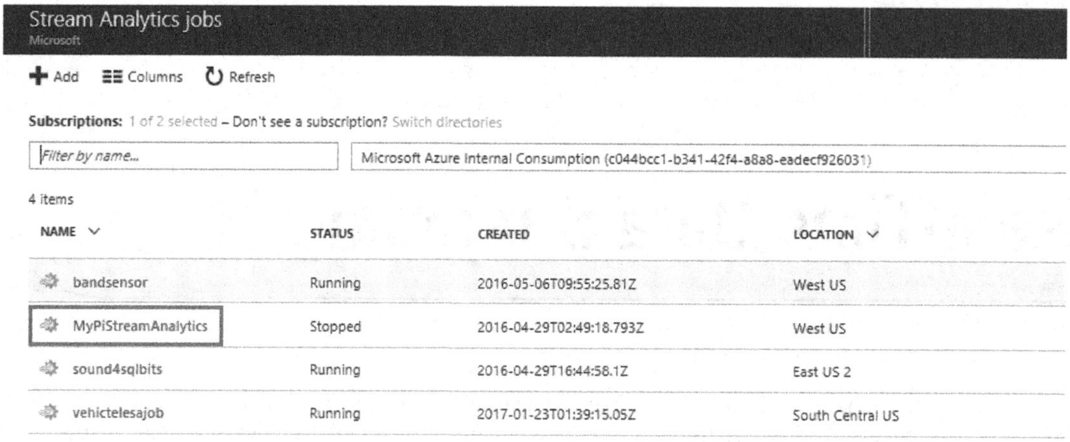

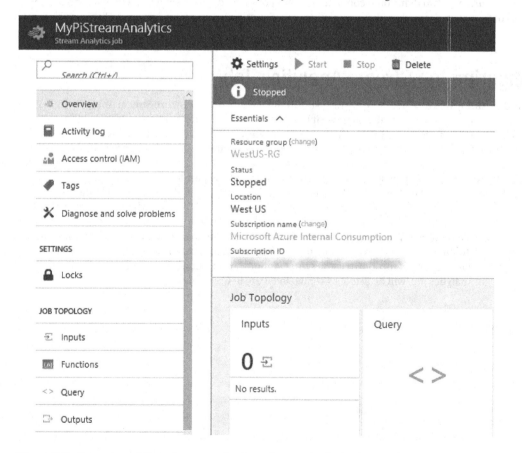

Figure 6-1. *Azure Stream Analytics jobs list*

Selecting the MyPiStreamAnalytics job from the list will open the Properties and Overview panes, as shown in Figure 6-2, from which the Stream Analytics job can now be configured.

Figure 6-2. *Property and Overview panes for Azure Stream Analytics*

The three steps needed to finish configuring the Stream Analytics job are creating an input, an output, and defining a query. Let's start by creating the input.

Creating an Input

The input will define where to pick up the data in order to begin the routing process. In the Overview pane, click the Inputs tile, which will open the Inputs pane. Any existing inputs will be listed in the pane showing the input alias name, source, and source type. Since this is a new Stream Analytics job, no inputs should exist, so click the +Add link on the Inputs pane.

In the New Input pane, provide an input name of iothubinput, and then configure the rest of the input as follows:

- **Source Type**: Data Stream

- **Source**: IoT hub

- **Subscription**: Use IoT hub from current subscription

- **IoT Hub**: MyPi3IoTHub

- **Endpoint**: Messaging

- **Shared Access Policy Name**: iothubownder

- **Consumer Group**: $Default

- **Event Serialization Format**: JSON

- **Encoding**: UTF-8

Once configured, the New Input pane should look like Figure 6-3. The IoT Hub is the hub created in Chapter 3, and a lot of the properties will have default values, such as the endpoint, consumer group, event serialization format, and encoding. However, it is good to ensure that these properties are set correctly before clicking the Create button.

Since the data coming in from the sensors and now sitting in Azure Hub is in JSON format, the input needs to configured to know which format to use for the incoming data. For encoding, only UTF-8 is currently supported. The Endpoint property has two values: Messaging and Operations Monitoring. When working with streaming messages from a device, which this example is showing, the appropriate value to select is Messaging. When working with telemetry and metadata, use the Operations Monitoring value.

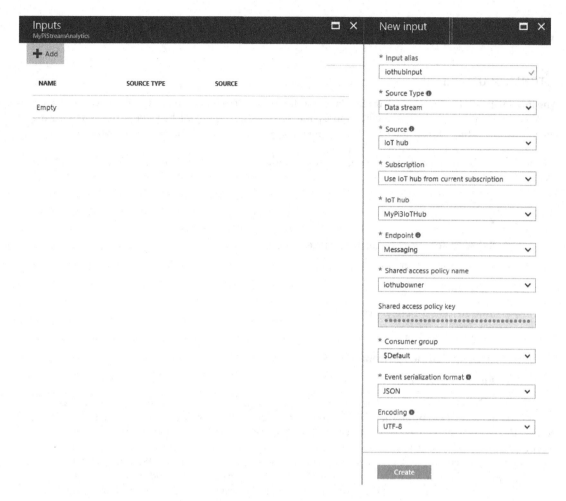

Figure 6-3. *Adding a new Azure Stream Analytics job input for IoT Hub*

There are a few things to discuss in more detail before creating the outputs, such as differentiating between data streams and reference data, and understanding consumer groups, so the next couple of sections will briefly discuss these concepts.

Data Stream vs. Reference Data

As seen in the Input wizard, there are two kinds of sources: data stream and reference data. Adding reference data as input is optional, but each Azure Stream Analytics job requires at least one data stream input. A data stream is a sequence of continuous, unbounded events arriving over a period of time from, as you saw, IoT Hub, Event Hub, or Blob Storage. In the example used in this book, IoT Hub was used because it is optimized to collect data from Internet-connected devices in IoT scenarios. Event Hub is similar but as will be discussed later in the book; there are some differences between IoT Hubs and Event Hubs.

Reference data is just what you would think it is: auxiliary or ancillary data that is static or changes very infrequently and is used as lookup or correlation data. Currently the only option for a reference data input is Azure Blob Storage.

Consumer Groups

Consumer groups allow you to virtualize a queue. They were popularized in their term and usage through Apache Kafka, which allows multiple consumers to consume their own "stream" of messages, managing the message pointer against the queue retention. This is especially useful in scenarios where systems are consumed by multiple external third parties and particularly within IoT where consumers are interested in different sliding window calculation periods of message content.

Creating an Output

The next step is to create the output. This example will be storing the data in Azure Data Lake Store, a hyper-scale data repository for big data analytic workloads. However, it doesn't exist yet, so let's go create it.

Azure Data Lake Store will be discussed in detail in the next chapter, so in this chapter you'll simply create it. In the portal, select New ➤ Intelligence + Analytics ➤ Data Lake Store, as shown in Figure 6-4.

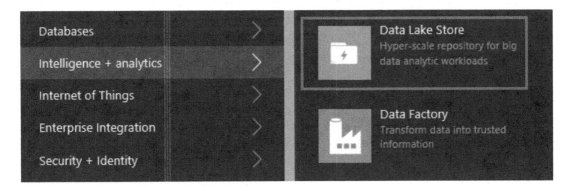

Figure 6-4. *Creating an Azure Data Lake Store account*

The New Data Lake Store blade opens up and, as you can see in Figure 6-5, there isn't a whole lot to enter or select to create a Data Lake Store. Simply give it a name, select the appropriate resource group and location, select the appropriate pricing option, and then click Create.

For pricing options, drop the list down to show the available pricing options. If you are not sure which is appropriate for your solution, click the information icon just above the Pricing option field, which will display an explanation and a link to read more information about the options.

The pay-as-you-go option charges a price per month based on the amount stored. The monthly commitment packages provide a significant discount up to 32% over the pay-as-you-go pricing. Click the link in the information pop-up for more information, but you will certainly want to calculate which option is the best for your solution.

As of this writing, the only locations available to create an Azure Data Lake Store are the East US 2 and Central US data centers. Many more data centers and regions will be coming online in the upcoming months and years.

In less than a minute the Azure Data Lake Store should be created and displaying the Overview and Settings blade.

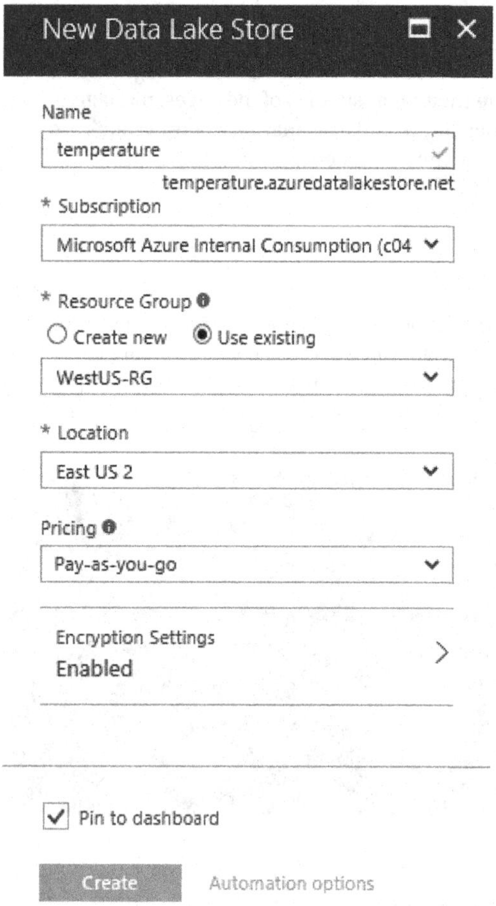

Figure 6-5. *New Data Lake Store blade*

As mentioned, the topic of Azure Data Lake Store will be covered in the next chapter, so at this point the next step is to configure the Azure Stream Analytics job output. Return to the Stream Analytics job either by clicking the tile on the main dashboard or searching as outlined earlier.

Once in the Stream Analytics job, you will be back at the Properties and Overview panes, as shown in Figure 6-2. This time, though, click the Outputs tile, which will open the Outputs pane. Any existing outputs will be listed in the pane, showing the output alias name and sync. Again, since this is a new Stream Analytics job, no outputs should exist, so click the +Add link on the Outputs pane.

In the New Output pane, the Sync property will automatically default to Event Hub, so change it to Data Lake Store. In the list options, as discussed, are hot path/cold path destinations. Hot path destinations include Power BI or Event Hub. Cold path destinations include SQL Database or even Blob Storage. Some on the list below could be both hot and cold path, depending on how long the data resides in the destination before processing. For example, Blob Storage and Data Lake Store could be both hot and cold. If the data is dropped in Blob Storage and processed quickly by Azure HDInsight, it could be considered a hot path for the data.

Select the Data Lake Store option, and you will notice immediately that pane quickly changes to what you see in Figure 6-6, which is asking you to authorize a connection to the Azure Data Lake Store created above. This authorization will grant Azure Stream Analytics permanent access to your Azure Data Lake Store account.

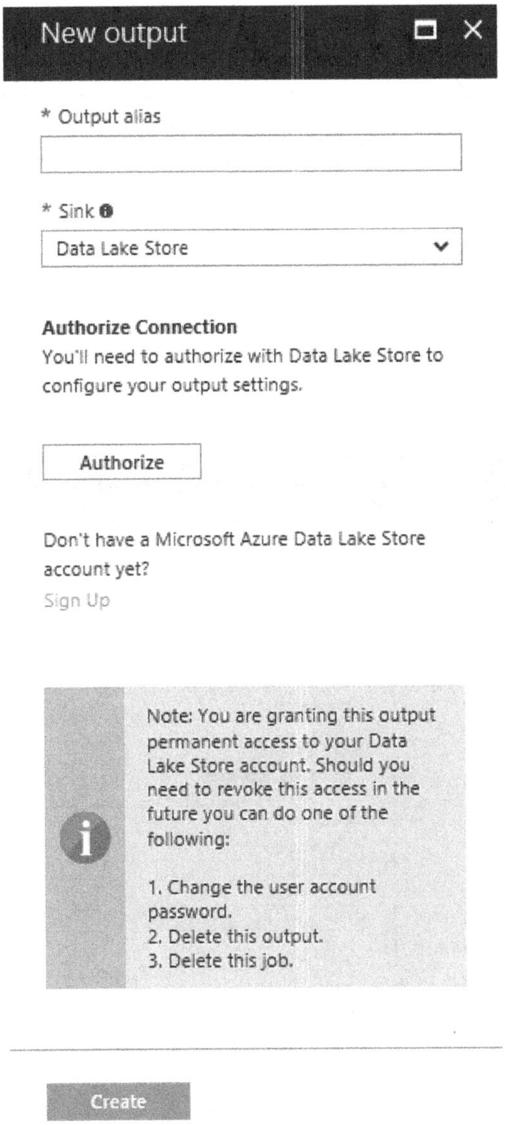

Figure 6-6. *Granting the output access to Azure Data Lake Store*

If for some reason the need arises to revoke access later, you can either delete the entire Stream Analytics job, delete this output, or change the account password. It is probably best to simply delete and recreate the output.

So, click the Authorize button and, when prompted, enter the appropriate authorization credentials, which are typically your Azure subscription credentials. It should be mentioned that this process generates an Azure Active Directory (AAD) token, but not a refresh token. If you authenticate with a Microsoft account here, you'll get a token for 48 hours, whereas if you authenticate with a work or school account, the token will last for two weeks.

Once the authorization is successful, the rest of the configuration options for the Azure Data Lake Store output will be shown in the New Output blade, as shown in Figure 6-7. As such, configure the rest of the input as follows:

- **Output Alias**: temperatureadls

- **Account Name**: temperature

- **Path Prefix Pattern**: tempcluster/logs/{date}

- **Date Format**: YYYY/MM/DD

- **Event Serialization Format**: JSON

- **Encoding**: UTF-8

- **Format**: Line separated

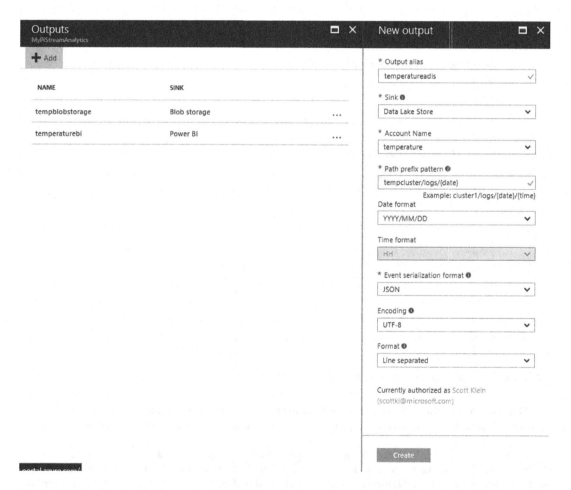

Figure 6-7. *Configuring the Stream Analytics job output*

The Path Prefix Pattern both sets and locates the location of the files coming into Data Lake Store. Together with the Date Format field, these two fields help organize and define the path and location of streaming data. To help with the organization there are two built-in variables available: {date} and {time}.

In Figure 6-7, a path of `tempcluster/logs` is specified with the variable {date} appended, and the date format of YYYY/MM/DD is used. As such, data will be stored in paths such as the following:

```
/tempcluster/logs/2016/06/15
/tempcluster/logs/2016/06/16
```

And so on.

It is possible to use the{date} and {time} variables more than once. Alternatively, there is no requirement that the variables must be used. This time granularity is used by Azure Data Factory (ADF), which is normally configured to pick up data every hour or so. If Azure Data Lake Store is used, like in this example, to store device data, then Azure Data Lake Analytics (ADLA) or HDInsight can be used through Azure Data Factory to process the data. Both Azure Data Factory, Azure Data Lake Analytics, and HDInsight are covered in later chapters.

For this output, notice that the output is saving the data in Azure Data Lake Store in JSON format. Currently there are three options for specifying the event output format: JSON, CSV, and Avro. Avro is an Apache data serialization format regularly used with Hadoop with support for a compact and fast binary data format, rich data structures, and the ability to store persistent data via a container file. Microsoft released the Microsoft Avro Library a couple of years ago and it can be installed via Nuget. This library can be used to serialize objects and other structures into streams such that they can be persisted to a data source such as a database or file.

As stated above, it is regularly used with Hadoop, and thus HDInsight, as a convenient way to represent complex data structures within a MapReduce job. The structure of an Avro file is such that it supports the MapReduce programming model, with the added benefit that the files can be split to allow the ability to seek any point in a file and start reading from a specific block. Read more about Apache Avro at `http://avro.apache.org/`.

For the format, there are two options; line separated and array. Line separated specifies that the output will be formatted with each JSON object separated by a new line. Array specifies that the object will be formatted as an array of JSON objects.

If you haven't already, click Create on the Create Output blade. Upon clicking Create, Stream Analytics will validate the connection with the Azure Data Lake Store, create the output, and add the output to the Stream Analytics job.

At this point, the input and output have been created (see Figure 6-8), so the next step is to write the query to "connect" the input and output to continue the movement of the sensor data.

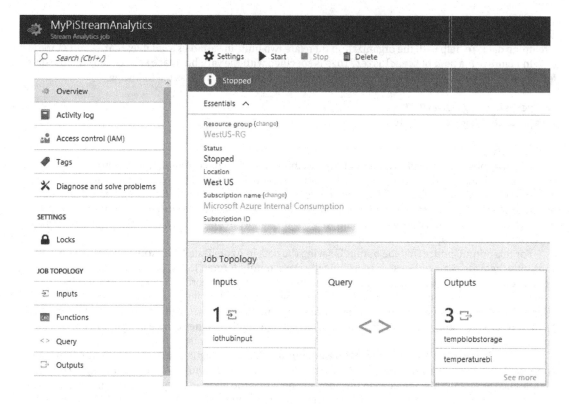

Figure 6-8. *Azure Stream Analytics job overview with input and output defined*

Defining the Query

The beauty of Azure Stream Analytics is the ease in which a Stream Analytics job can be created with just a few clicks, the result of which is real-time event processing. The core of this process is in the data transformation ability, which is expressed in a SQL-like language. Once the input and output are defined, the real power and attractiveness of ASA is in the declarative query model for describing transformations, removing the need of dealing with complex stream processing systems. If you know T-SQL, you can write Stream Analytics queries with ease.

The SQL-like query language is a subset of the T-SQL syntax, offering many of the same similar keywords, including SELECT, FROM, WHERE, GROUP BY, CASE, HAVING, JOIN, and much more.

The basic query is a simple pass-through query which picks up the data from an input and passes it through to an output, as shown in Figure 6-9.

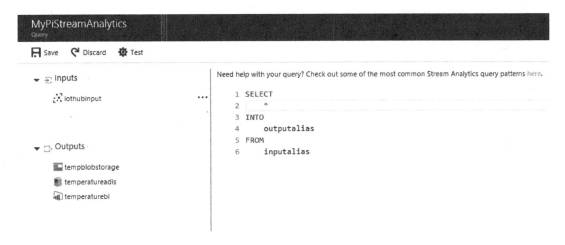

Figure 6-9. *Simple pass-through query*

This query simply says "select everything from the input alias and insert into the output alias." Thus, by replacing the default values in the brackets with the input and output aliases created earlier in this chapter, the query now looks like the following:

```
SELECT
    *
INTO
    temperatureadls
FROM
    iothubinput
```

Save the query by clicking the Save link. The Stream Analytics job is finished and ready to be run.

What this is doing, as explained above, is simply moving all incoming data from Azure IoT Hubs to Azure Data Lake Store. Before running the job, let's discuss the possibilities here. While in this example there is only a single sensor, in this case a single Raspberry Pi, what if there were multiple sensors? In my environment, I have a Raspberry Pi 3 with a DHT22 sensor, a Raspberry Pi 2 with a DHT11 sensor, the Tessel 2 with its Climate module, and a Raspberry Pi3 with a FEZ HAT attached to it. All four are sending data to Azure IoT Hubs, and the query above is sending all the data into Azure Data Lake Store. What if there were hundreds or thousands of sensors? Maybe you would want to be more specific about where data was sent. Maybe some of the data would be sent to Azure Blob Storage or Azure SQL Database or event Azure Event Hubs based on their location or type of sensor for different processing. These are things to certainly think about and in the "More on Queries" section below I'll discuss how queries can help clean or filter data based on conditions.

Since we are talking about queries here, I'll also mention that it is possible to have multiple queries in the query window to send the data to multiple outputs. For example, it is possible to do the following within the Stream Analytics job:

```
SELECT
    *
INTO
    temperatureadls
FROM
    Iothubinput

SELECT
    *
INTO
    tempblobstorage
FROM
    Iothubinput
```

Thus, when the job runs, the Stream Analytics job will select the data from the defined input and route the data to both outputs. In other words, both outputs will get the same data.

Running the Stream Analytics Job

Now you are at the point where you can start the Stream Analytics job, turn on the devices, and start seeing data routed through Azure IoT Hubs, picked up by Azure Stream Analytics, and dropped in Azure Data Lake Store.

Start the Stream Analytics job by click the Start button at the bottom of the portal. A pop-up will ask when the job will start generating output. By default, Job Start Time is selected, which is fine. Click OK. The Stream Analytics job will start processing data and generating output. However, there is no data to process, so you need to go back and turn on the devices to start sending data.

Go back to the application created in Chapter 2 and run it, ensuring that everything is still connected properly, the Raspberry Pi is on, and everything is ready to go. If all is well in the land of OZ, data will start being generated via the Raspberry Pi and the temperature sensor.

After a few minutes, you should be able to see activity in Azure IoT Hubs and Azure Stream Analytics via the Monitoring graphs in both blades. To really verify that all is working well, let's go to three places in the portal. First, head over to the IoT Hub; on the Overview pane you can see that messages are indeed being received, as seen in Figure 6-10.

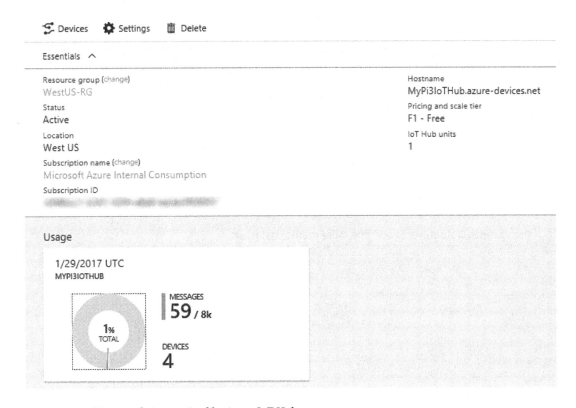

Figure 6-10. *Messages being received by Azure IoT Hub*

Next, head back over to the Azure Stream Analytics Job Overview pane, and sure enough, in the Monitoring graph you will see events being processed, as seen in Figure 6-11.

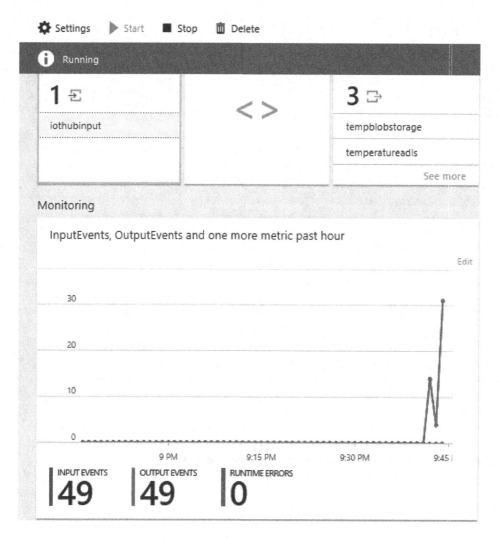

Figure 6-11. *Messages processed by Azure Stream Analytics*

Lastly, to validate that the Stream Analytics job output is indeed sending data to the Azure Data Lake Store, head over to the Azure Data Lake Store account, and on the Overview pane, click the Data Explorer link.

In the left navigation pane, drill down in the tempcluster folder. Remember when the output was created, the Path prefix pattern was defined as tempcluster/logs/{date}. Thus, drilling into the folder structure, you will see the logs subfolder followed by the year and month subfolders, as shown in Figure 6-12. And sure enough, in the month folder is a file containing the temperature data that was inserted by the Stream Analytics Job output.

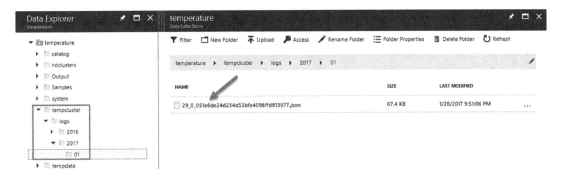

Figure 6-12. *Data inserted into Azure Data Lake Store*

You can view the contents of the file by simply clicking the file, which will open the File Preview pane. You can also download the file by clicking the Download button while in the File Preview pane, or back in the file listing, right-clicking the file and selecting Download from the context menu.

The exercise in this chapter used Stream Analytics to stream data from Azure IoT Hubs to Azure Data Lake Store. The query wasn't too complicated and it was fairly easy to set up Azure Data Lake Store and configure a Stream Analytics output to it.

The next section will discuss queries in Stream Analytics in more detail, but there is also more that can be done with outputs. Another interesting output is Power BI. This is where the meat of real-time analytics comes into play. There is a whole chapter dedicated to examples using Stream Analytics outputs to Power BI later in the book. This is quite powerful so you won't hurt my feelings if you jump to that chapter right now. Just make sure you come back and pick up where you left off. ☺

There are other interesting outputs, such as Azure SQL Database or DocumentDB. DocumentDB is a full-managed NoSQL document data database service. Document DB natively supports JSON documents, which makes it a great candidate for cold or warm path scenarios if you are routing events in JSON format. Azure SQL Database is a relational database-as-a-service. It is SQL Server running as a service.

While this chapter won't specifically walk through creating additional outputs, this chapter will hand out a homework assignment. Lucky you. The homework assignment is to create an additional output to either Azure Blob Storage, Azure SQL Database, or DocumentDB. You should notice in Figures 6-8 and 6-11 that I have already created an output to Azure Blob Storage. As an FYI, the Event Serialization Format property was set to CSV for the Azure Blob Storage output.

As an additional FYI, the Stream Analytics job will need to be stopped in order to modify it. In other words, to create new inputs or outputs, or to modify the query on a specific Stream Analytics job, the job must be stopped.

Therefore, stop the job and add an additional output to either Azure Blob Storage, DocumentDB, or Azure SQL Database. The next section on queries will show how to modify the query to send data to both Data Lake Store as well as another output within the same query.

More on Queries

Before moving on, it will be beneficial to spend a bit more time on the query syntax for Stream Analytics and cover some of the SA-specific items. Honestly, there really isn't that much to it. As stated earlier, the SQL-like query language is a subset of the T-SQL syntax, which makes writing queries quite simple.

In its simplest form, the basic query is a simple pass-through query:

```
SELECT
    *
INTO
    outputalias
FROM
    inputalias
```

Filtering can be applied based on a condition. The following is only passing through the events for the Pi3 device via the WHERE clause:

```
SELECT
    Device AS DeviceName,
    Temp AS Temperature,
    Humidity AS Humidity,
    Time AS ReadingTime
INTO
    outputalias
FROM
    Inputalias
WHERE Device = 'Pi3'
```

The following query adds another common T-SQL clause, HAVING, but also introduces a Windowing function specific to ASA, the TumblingWindow. This example does not filter on a specific sensor, rather it monitors the average temperature over 30-second windows and only passes through events if the average temperature is over 90 degrees:

```
SELECT
    Device AS DeviceName,
    Time AS ReadingTime,
    Avg(Temp) AS AvgTemperature
INTO
    outputalias
FROM
    Inputalias TIMESTAMP BY Time
GROUP BY TumblingWindow(second, 30), Device
HAVING Avg(Temp) > 90
```

As is seen in the examples above, Stream Analytics makes it quite easy to perform streaming computations over streams of events by offering a nice and comprehensive subset of T-SQL syntax. The only caveat to this is understanding how Windowing functions work, so this chapter will wrap up by taking a look at the Windowing functions in ASA.

Windowing

Many times when processing real-time incoming events, it is necessary to aggregate subsets of events in which these events fall within a specified period of time. For example, to show the average temperatures in 10-second windows or sum values of incoming events in 5-second windows.

To make this easy in Stream Analytics, subsets of events are defined through what is known as windows, which represents a grouping of time. Within a window is contained event data along a timeline, and Stream Analytics via the window allows you to perform different operations against the events within that window.

There are three types of window functions: tumbling, hopping, and sliding. The general syntax for the three Windowing functions is the following:

```
<functionname>(timeunit, windowsize)
```

The timeunit is the unit of time for the windowsize. The following table lists the valid arguments for the timeunit argument:

Timeunit	Abbreviation
day	dd, d
hour	Hh
minute	mi, n
second	ss, s
millisecond	Ms
microsecond	Mcs

The windowsize argument is a big integer that specifies the size of the window. For example,

```
TumblingWindow(second, 30)
```

The lowest or smallest unit of time that can be used in a windowing function is 1 second. Related to that, one question that is asked somewhat frequently pertains to ASA performance when using a windowing function in a query. While there are no specific numbers or documentation describing Azure Stream Analytics query performance, it is true that windowing functions typically require ASA to hold some event values for a period of time. This results in a higher resource utilization. Therefore, the bigger the window size and the larger the number of events, the more resources are needed. The following sections discuss the specifics of each Windowing function.

Tumbling Window

The tumbling window is a fixed-size series of contiguous time intervals, and these time intervals do not overlap. One thing to note, however, is that both this function and the hopping window function are inclusive in the end of the window and exclusive in the beginning. Meaning, a 12:00pm – 1:00pm window will include events that happen exactly at 1:00pm but not include events that happen at 12:00pm. The offset argument swaps the behavior to include the events at the beginning of the window and excludes the events at the end.

The syntax for the tumbling window is simply the following:

```
TumblingWindow(timeunit, windowsize, [offset])
```

For example,

```
TumblingWindow(second, 5)
```

You will see an example of a TumbingWindow in Chapter 13 in which Azure Stream Analytics will send data to a Power BI output. When running through the example in Chapter 13, feel free to include two queries to send data both to Power BI and another output, such as Azure Blog Storage, so that you can download the data and get an idea of what the output looks like.

Hopping Window

Hopping windows differ from tumbling windows in that they are meant to overlap. The general syntax is as follows:

```
HoppingWindow(timeunit, windowsize, hopsize, [offset])
```

The third parameter in the function specifies the hopsize, or, the window overlap time (in other words, the amount of time each window moves forward relative to the previous window). Thus, an example of a hopping window is

```
HoppingWindow(seconds, 10, 5)
```

In this example, the window is 10 seconds in length, but the hop is 5 seconds, so the latest window overlaps with the previous window by seconds.

Sliding Window

Sliding windows operate a bit differently than the other two windows. Stream Analytics logically manages and looks at all windows of a given length and only outputs events when an event has entered or exited the window. Every window will have at least one event and the window continuously moves forward by an Epsilon.

The general syntax for the SlidingWindows function is as follows:

```
SlidingWindow(timeunit, windowsize)
```

Thus, the function might be used like so:

```
SlidingWindow(second, 5)
```

A popular example of this query is the following, which says "Give me the count of tweets for all topics that are tweeted more than 10 times in the last 10 seconds:"

```
SELECT Topic, COUNT(*)
FROM TwitterStream
TIMESTAMP BY CreatedAt
GROUP BY Topic, SLIDINGWINDOW(second, 10)
HAVING COUNT(*) > 10
```

A common scenario is one that takes the incoming JSON with hierarchies and turns it into CSV format via the query. This scenario, as is the incoming JSON, is a bit more complex but the solution to this is to use the "." syntax to access nested fields. For example, if your data looks like {a:1, b:{x:2, y:3}}, you can use a query such as the following to flatten it:

```
SELECT a, b.x, b.y FROM input
```

Window functions might take a bit of time to get used to, but as you will see in Chapter 13, they have great value for real-time data analysis. Visit `https://docs.microsoft.com/en-us/azure/stream-analytics/stream-analytics-window-functions` for an introduction to Window functions.

Summary

This chapter continued the example from Chapter 4, connecting the Azure IoT Hubs created in Chapter 4 with the Azure Stream Analytics created in Chapter 5. You created the necessary input to pick the events up from Azure IoT Hubs, the output to drop the events into Azure Data Lake Store, and the query to execute and orchestrate the stream.

This chapter also spent some time discussing and looking at queries and the similarities to T-SQL, and then drilled into the Stream Analytics windowing functions and how they can be used to aggregate subsets of events to gain valuable insight into events within a specific timeline.

CHAPTER 7

■ ■ ■

Azure Data Factory

So far in your example, you have data sitting in both Azure Data Lake Store and Azure Blob Storage, waiting for the next step in the journey. Chapters 5 and 6 walked through the process of using Azure Stream Analytics to pick the data up from Azure IoT Hub and route it to hot path or cold path destinations, depending on the analysis needs for data insights.

Revisiting the hot path/cold path scenario for a moment, your data has followed a cold path, meaning it is sitting in storage for prompt processing and analytics by other services. Chapter 13 will look at and discuss hot path scenarios, specifically sending data to Power BI for immediate analysis.

However, let's take a step back. What you have done so far is produce data (non-relational data, to be specific) from a small handful of devices and route it to a storage medium in the cloud. That's what you intended, but let's step back even further and revisit the bigger picture discussed in Chapter 1. What you have covered so far is a small, but important, section of the bigger picture wherein you are consuming data at a rapid pace and preparing it for further analysis.

In any business, though, the data you have generated so far is only a fraction of the data needed to make intelligent and timely business decisions. Looking at it purely for what it really is, you have simple temperature data. That's it. You know when the readings were generated and from what device, but that's it. But businesses have much more data that can provide additional insight and play a supporting role for the streaming data coming in. For example, some professional sports teams are using Azure Data Factory to help optimize marketing campaigns to drive optimum engagements with their global fan base. The clubs and teams can better understand how users are interacting with their apps and websites for better monitoring and insights.

Thus, several questions come to mind. How is existing data leveraged in business? Is it possible to enrich streaming data by leverage existing reference data from additional on-premises data sources or other disparate data sources? In the world of big data and IoT, what is needed is a platform for aggregating and processing data from a wide variety of sources.

Azure Data Factory is such a service, and this chapter will take a detailed look into this critical service, from what it is, how to create and configure it, the scenarios in which it can be used, and how it differs from other data aggregation and processing technologies.

What Is Azure Data Factory?

Working across both on-premises and the cloud, Azure Data Factory is a data integration service that is designed specifically to collaborate with existing services for the movement, transformation, and processing of raw data from disparate systems and transform it into useful information.

The key words in the paragraph above are "data integration," in that Azure Data Factory is an enterprise platform in which reliable, repeatable, and deterministic data movement and transformation flows are architected and built.

© Scott Klein 2017
S. Klein, *IoT Solutions in Microsoft's Azure IoT Suite*, DOI 10.1007/978-1-4842-2143-3_7

On the surface, Azure Data Factory might seem like a simple data ingestion and transformation engine. The truth is far from that. In reality, Azure Data Factory is an enterprise-level service that allows you to plug in to a plethora of supporting technologies and services to achieve the deep, rich data insight and analysis that normal ETL technologies simply cannot provide. For example, you can use Azure Data Factory to kick of a Hadoop/HDInsight job for big data analysis each morning, or push the data to Azure Machine Learning for a deep analysis of trends and behaviors in your data monthly. The possibilities are nearly endless.

Azure Data Factory enables these fundamental capabilities through the use of key concepts and components, discussed next.

Key Components

Azure Data Factory is composed of four key components that work in tandem to provide the collection, aggregation, and processing discuss above. Together they provide the platform in which to compose simple to complex data movement and transformation orchestrations for your data flow.

Activity

An *activity* is a collection of actions in which to perform on your data, grouped together in a unit of orchestration and execution. Each activity can have zero or more input datasets and produce one or more output datasets. For example, an activity could be copying data from one dataset to another, or calling a stored procedure in SQL Server, Azure SQL Database, or Azure SQL Data Warehouse.

Pipeline

A *pipeline* is a group of *activities*. Together, the activities in a pipeline perform a task. For example, a pipeline could contain a set of activities that ingest and clean log data, and then kick off a HIVE query on an HDInsight cluster to analyze the log data. The beauty of this is that the pipeline allows you to manage the activities as a set instead of each one individually. For example, you can deploy and schedule the pipeline, instead of the activities independently.

Dataset

A *dataset* is a named view of data that simply points or references the data you want to use in your *activities* as inputs and outputs.

Linked Service

You can think of a *linked service* much like a connection string. A linked service simply defines the connection information in which to connect to external data sources. Think of it this way: the *dataset* represents the structure of the data, and the *linked service* defines the connection to the data source. There is one minor caveat to this but this caveat will be discussed when discussing linked services in detail later in the chapter.

With that background and knowledge, it is time to walk through how to create and configure an Azure Data Factory. This information will come in handy in Chapter 8 when you create an Azure Data Factory to continue the example from Chapter 6. The following section will walk through the creation of a Data Factory and discuss the different configurations and settings available. This chapter will show you how to do it using both the portal and Visual Studio.

Creating and Configuring a Data Factory

Currently there are four ways to create an Azure Data Factory: using the Azure portal, using Visual Studio, using PowerShell, or with the Resource Manager Templates. The next two sections will illustrate how to create and configure an Azure Data Factory in preparation for creating an Azure Data Factory in the next chapter to continue the example from Chapter 6. These sections will use both the portal and Visual Studio since both provide excellent GUIs for managing and maintaining pipelines and other aspects. Let's begin with the portal.

Portal

Creating the Azure Data Factory is actually pretty easy. If you still have the portal open in the browser, click New ➤ Data + Analytics and select Data Factory, as shown in Figure 7-1. If you don't have the portal open, navigate to Portal.Azure.com and log in. Then follow the instructions above.

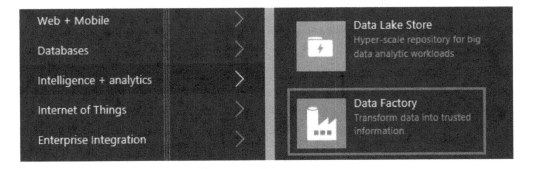

Figure 7-1. *Creating an Azure Data Factory*

When the New Data Factory blade opens, simply provide a name for the Data Factory, and then select the subscription, Resource Group, and Location, as shown in Figure 7-2.

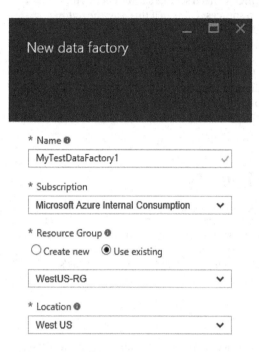

Figure 7-2. *Configuring the Data Factory*

Make sure the Resource Group and Location are in the same region/data center. Also, select the "Pin to dashboard" checkbox and then click Create.

The Data Factory will take less than a minute to create, at which point you will be presented with the Overview and Settings blade. The Overview blade is shown in Figure 7-3.

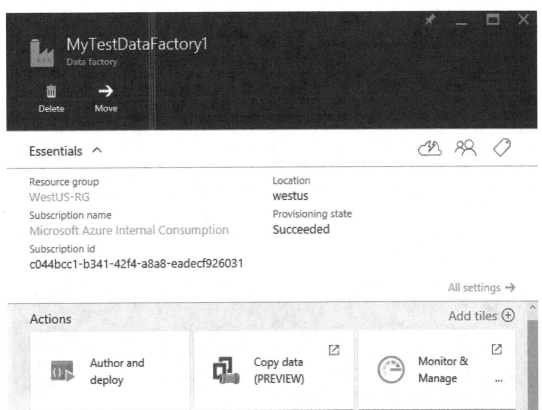

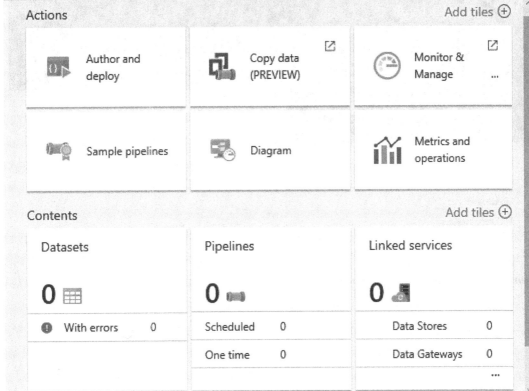

Figure 7-3. *The Azure Data Factory Overview pane*

The Settings blade won't be covered here simply because it is very similar to other Settings blades discussed so far with other services. The Settings blade allows you to access the audit logs, provide tags, and manage users, which you have seen previously.

The important pieces to Azure Data Factory are in the Actions section of the MyTestDataFactory1 blade, and this is where the majority of the time is spent in configuring a Data Factory. At this point, the Data Factory has been created, but nothing has been defined (such as an activity or pipeline), nor has anything been deployed.

Taking a good look at this blade, there are key links in the Actions section of the blade. The Author and Deploy tile is where activities and pipelines are defined and orchestrations are built. The Copy Data tile opens a new web page that provides a wizard for creating a simple pipeline for copying data from supported sources to destinations. The Monitor & Manage tile also opens up a new web page, this one containing an application that provides insight into pipeline resources, sets and views alerts, and monitors activities in the Data Factory. The Diagram tile provides a visual representation of pipelines (you'll see an example of this in the next chapter).

In the Actions section of the MyTestDataFactory1 blade, click the Author and Deploy link. This will open up the blades with which to define the components of the Data Factory, shown in Figure 7-4. Essentially, this is where the unit of orchestration is built.

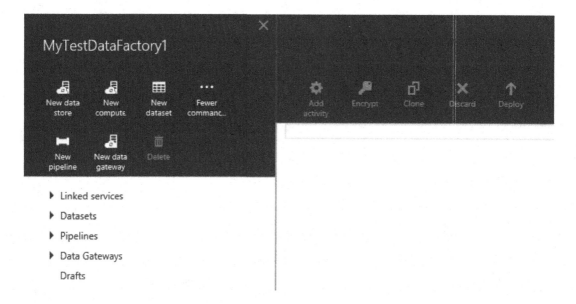

Figure 7-4. *Authoring an orchestration*

The key to constructing a pipeline orchestration is two-fold; first, knowing in which order to create the components and second, knowing how to write JSON. Tackling the first issue, Figure 7-4 shows buttons for creating new data stores, new compute, new datasets, and new pipelines. Directly beneath the buttons are the lists of created components, such as linked services, datasets, etc. So where is the "New Linked service" button? Where is the "New Activity" button? The confusion is mapping the buttons to the components found in the Azure Data Factory documentation, which was discussed above.

To clear up the confusion, this section will walk through a simple example of creating a pipeline. Now, in terms of the second issue, JSON, luckily a deep knowledge of JSON isn't necessary since the portal provides nice templates. All you need to do is fill in the necessary information.

The first step in creating a pipeline orchestration is to create the linked services. This is done by clicking the "New data store" button, which presents a nice list of 18 data stores to connect to. Take a moment to explore the list, in which you'll find data stores that include the file system, ODBC, OData, relational stores such as Sybase, Oracle, DB2, SQL Server, and SQL DB, and non-relational stores such as Azure Storage and HDFS. It's quite a list.

Selecting any of the data stores populates the right side of the blade, the Draft blade, with a JSON template specific to that data store. The great thing about the data store templates is that they basically need the same information: a name, a type, and a connection string. The type is filled in by default since you are picking the data store type, and the name is also filled in by default. For example, the following is the template for the Azure Storage data store:

```
{
    "name": "AzureStorageLinkedService",
    "properties": {
        "type": "AzureStorage",
        "description": "",
        "typeProperties": {
            "connectionString": "DefaultEndpointsProtocol=https;AccountName=<accountname>;
            AccountKey=<accountkey>"
        }
    }
}
```

Notice that the name contains "LinkedService." Thus, creating either a new data store or new compute creates a linked service. As discussed above, a linked service is essentially a connection string to a data source. Thus, the difference between a *data store* and a *compute* is that the data store represents storage of relational or semi-relational data while compute represents resources that can host the execution of an activity, such as HDInsight or Azure ML. Chapter 12 discusses Azure HDInsight and Chapter 14 covers Azure ML (Machine Learning).

Additionally, when creating a data store linked service, depending on the data store, typically all you need to fill in is the connection string information. In the example above, really all you need to fill in is the account name and account key.

Take a few minutes and click the different data store types to see what kind of connection information they need. An interesting data store is HDFS, which is the Hadoop Distributed File System, the built-in file system for Hadoop clusters. The JSON connection requires additional information but the format is still similar to the others. For a compute linked service, the connection requires a bit more information, and you will see an example of that shortly.

To show how this works, I have created a new Azure storage account called scottkleinw2 and then created a Data Factory Azure Storage data store. I filled in the connection information (AccountName and AccountKey) and then clicked Deploy. I now have an Azure Storage linked service, as shown in Figure 7-5.

Figure 7-5. *Creating an Azure Storage linked service*

In regards to security, how are the credentials stored when creating a linked service? Let's take the recent of scenario, shown in Figure 7-5, as an example. The information necessary to access the Azure Storage account is entered as the connection string, including the account name and account key. When the linked service is deployed, the credentials are encrypted and stored internally in Azure Data Factory and are only available to Azure Data Factory. The same applies to any other linked service connection information, such as Azure SQL Database.

Back to this example, this is a linked service to a data store where the data currently resides, awaiting some type of process or activity on the data. Thus, the next step is to create a new linked service that will do some type of execution on that data. So, click the New compute button to create a new compute linked service. And explore the different types of compute linked services. There are only a handful, including HDInsight (on-demand or existing), Azure Batch, Azure ML (Machine Learning), and Azure Data Lake Analytics (ADLA).

For this example, I want to create a new on-demand HDInsight cluster. This will automatically create an HDInsight cluster at runtime and delete it when the process is done after a specific amount of idle time. Go ahead a select the on-demand HDInsight cluster and notice that it follows the same type of pattern: a name, type, and cluster information. The difference here is that since it creates it at runtime, you only need to specify the size of the cluster, cluster type (Windows or Linux), version, idle time, and the storage linked service name that you created earlier. The Azure Storage linked service is used to store the resulting data from HDInsight processing (i.e. logs from a Hive job). You have the option to store these logs on the same storage you're pulling data from or a different account.

Configure the cluster by replacing the properties with the following values:

- **clusterSize**: 2

- **timeToLive**: 00:02:00

- **osType**: Windows

- **version**: 3.4

- **linkedServiceName**: AzureStorageLinkedService

Click Deploy to save and deploy the linked service. Click the new HDInsight linked service to show the following:

```
{
    "name": "HDInsightOnDemandLinkedService",
    "properties": {
        "description": "",
        "hubName": "mytestdatafactory1_hub",
        "type": "HDInsightOnDemand",
        "typeProperties": {
            "version": "3.4",
            "clusterSize": 2,
            "timeToLive": "00:02:00",
            "osType": "Windows",
            "coreConfiguration": {},
            "hBaseConfiguration": {},
            "hdfsConfiguration": {},
            "hiveConfiguration": {},
            "mapReduceConfiguration": {},
            "oozieConfiguration": {},
            "sparkConfiguration": {},
            "stormConfiguration": {},
            "yarnConfiguration": {},
            "additionalLinkedServiceNames": [],
            "linkedServiceName": "AzureStorageLinkedService"
        }
    }
}
```

This code is quite interesting because it shows the different cluster configuration properties available for HDInsight. These properties are not required, but specify additional granular configuration for the on-demand HDInsight cluster linked service.

In the example so far, Azure Storage data store and On-demand HDInsight cluster linked services have been created, but there is still a bit more to do. The next step in creating a pipeline is to create the datasets. Typically, a pipeline will have an input and output dataset. The input dataset will specify the source data (i.e. the location of the data). Your linked service simply points to the Blob Storage account; the dataset will specify what container/folder and file(s) to process. The output dataset will specify the container/folder in which to drop the results of the activity.

Click the New dataset option in the portal, and select Azure Blob Storage. The JSON for this will look quite complex and overwhelming, but it really isn't too bad once you get a good look at the contents. If you plan on partitioning the data (for example if you want to separate the data into different folders based on month/day/year), the JSON is preformatted to help you easily supply the necessary information.

For this example, you're not going to do that since this will be the input dataset so the JSON for this will be the following, which simply points to the container/folder and filename that the HDInsight cluster will pick its data up from, the text type and text delimiter, and the frequency in which the data input slices will be available. The external property specifies that the data is not being generated within the Data Factory service.

```
{
    "name": "AzureBlobInput",
    "properties": {
        "type": "AzureBlob",
        "linkedServiceName": "AzureStorageLinkedService",
        "typeProperties": {
            "fileName": "data.csv",
            "folderPath": "adfsample/input",
            "format": {
                "type": "TextFormat",
                "columnDelimiter": ","
            }
        },
        "availability": {
            "frequency": "Month",
            "interval": 1
        },
        "external": true,
        "policy": {}
    }
}
```

You'll deploy the input dataset, and then repeat the process to create the output dataset, this time changing the folderPath property to specify the output folder, and eliminating the external property since it isn't needed here, and deploy the output dataset.

```
{
  "name": "AzureBlobOutput",
  "properties": {
    "type": "AzureBlob",
    "linkedServiceName": "AzureStorageLinkedService",
    "typeProperties": {
      "folderPath": "adfsample/output",
      "format": {
        "type": "TextFormat",
        "columnDelimiter": ","
      }
    },
    "availability": {
      "frequency": "Month",
      "interval": 1
    }
  }
}
```

A quick comment on delimiters. The preceding example references a text file (CSV file) with columns delimited by a comma. Azure Data Factory supports any delimiter, including custom delimiters, as long as the delimiter is a single character.

You're done with the datasets, so the last step is to create the actual pipeline by clicking New pipeline. It is here in the pipeline that the activities are actually defined. As shown in the code below, an activity of type

HDInsightHive is being created and executed, meaning that when the pipeline executes, the on-demand HDInsight cluster will be created and the specified HIVE queries specified in gethightemp.hql file of the scriptPath property will be executed.

The activity will use the datasets specified in the inputs and outputs properties for its input and output. The defines section specifies the runtime settings that will be passed to the HIVE script as configuration values.

```
{
    "name": "TestPipeline",
    "properties": {
        "description": "",
        "activities": [
            {
                "type": "HDInsightHive",
                "typeProperties": {
                    "scriptPath": "adfsample/gethightemp.hql",
                    "scriptLinkedService": "AzureStorageLinkedService",
                    "defines": {
                        "inputtable": "wasb://adfsample@scottkleinw2.blob.core.windows.net/
                        input",
                        "outputtable": "wasb://adfsample@scottkleinw2.blob.core.windows.net.
                        blob.core.windows.net/ouput"
                    }
                },
                "inputs": [
                    {
                        "name": "AzureBlobInput"
                    }
                ],
                "outputs": [
                    {
                        "name": "AzureBlobOutput"
                    }
                ],
                "policy": {
                    "concurrency": 1,
                    "retry": 3
                },
                "scheduler": {
                    "frequency": "Month",
                    "interval": 1
                },
                "name": "RunHiveActivity",
                "linkedServiceName": "HDInsightOnDemandLinkedService"
            }
        ],
        "start": "2016-04-01T00:00:00Z",
        "end": "2016-04-02T00:00:00Z",
        "isPaused": false,
        "hubName": "mytestdatafactory1_hub",
        "pipelineMode": "Scheduled"
    }
}
```

Keep in mind that an activity does not need or require an input, but for the purposes of this example and to visual the diagram of a pipeline, an input was included.

Once the pipeline has been configured and published, the work of creating the pipeline orchestration is done. Going back to the MyTestDataFactory1 main blade and clicking the Diagram tile opens up the Diagram blade in which an overview of the pipeline and its datasets is displayed, as shown in Figure 7-6.

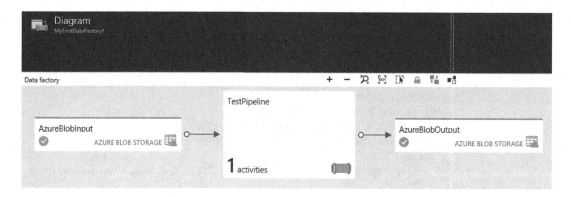

Figure 7-6. *Completed pipeline diagram*

Notice that the pipeline shows one activity, which is correct. Right-clicking the activity and selecting Open pipeline in the context menu will open a diagram in which all the specified activities and their inputs are displayed.

Monitoring of the Data Factory is done in a couple of places. While in the diagram, double-clicking any of the datasets opens up a summary of the dataset, which also includes a Monitoring tile that shows updated data slices. Clicking any of the data slices allows you to drill into the run of the slice and troubleshoot any errors and issues.

Monitoring of the activities is done by clicking the Monitor and Manage tile on the Data Factory blade. This opens up a new web page that also displays the pipeline but allows you to drill into the activity itself, viewing all executions of the activity, status, and any errors. Chapter 8 will take a look at the Monitoring and Management aspects of a pipeline.

Now, this was a very simple example to introduce you to Azure Data Factory and how to create a Data Factory, explore the key components of a Data Factory, and see the overall look and feel of Azure Data Factory. The next section will show how to do the same thing using Visual Studio.

Visual Studio

It is awesome to see Microsoft embracing Visual Studio as a means for working with many of the analytic and big data Azure services. The whole goal in doing this was to make the authoring, management, and deployment experience much easier and richer. Visual Studio is a very rich development environment in terms of productivity and efficiency, and so the benefits of being able to interactively author and deploy Data Factory solutions are undeniable, including integration with the Diagram view, Server Explorer, and the rich IntelliSense for editing JSON.

To get started, you first need to download and install the Azure Data Factory Tools for Visual Studio, found here:

`https://visualstudiogallery.msdn.microsoft.com/371a4cf9-0093-40fa-b7dd-be3c74f49005`

Once installed, create a new Data Factory project by selecting the DataFactory template, and then select the Empty Data Factory Project, shown in Figure 7-7.

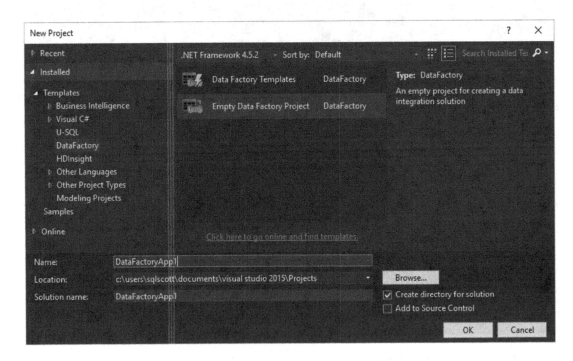

Figure 7-7. *Creating a Data Factory Project in Visual Studio*

Creating a Data Factory pipeline follows the same process as creating a Data Factory in the portal: you first create the linked services, then the datasets, then the pipeline.

Once the project is created, open up Solution Explorer and notice that the solution structure almost follows the same structure found in the portal, meaning you see linked services, pipelines, and tables. A table in Azure Data Factory is simply a named view of the data. Think of a table as a dataset.

For some reason, the Data Factory tooling team for Visual Studio didn't quite follow the same naming conventions as the portal team, so in Visual Studio, *datasets* are called *tables*. That's ok because they do the same thing and the JSON is exactly the same, which means, to follow this example, you can simply copy and paste from the portal into Visual studio. Awesome.

So, right-click linked services and select Add ➤ New Item. In the Add New Item dialog, select "Azure Storage Linked Service," as shown in Figure 7-8, and then click Add.

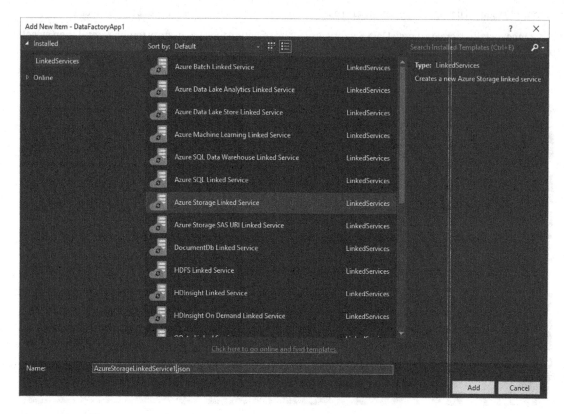

Figure 7-8. *Adding an Azure Storage linked service to the project*

Notice that the JSON template here is exactly the same template you saw earlier in the portal example. Thus, you can simply copy the JSON code from the AzureStorageLinkedService from the portal and paste it in Visual Studio, and then save it.

Create another LinkedService, this time selecting the "HDInsight On Demand Linked Service" template. Copy the code from the portal for the HDInsightOnDemandLinkedService, paste it here, and save it.

For the datasets, right-click the Tables folder in Solution Explorer and select Add ➤ New Item, and in the Add New Item dialog, click the Azure Blob template and click Add. Copy the code from the portal for the AzureBlobInput dataset and paste it into Visual Studio. Repeat this process for the AzureBlobOutput.

Lastly, right-click the Pipelines folder in Solution Explorer and select Add ➤ New Item, and in the Add New Item dialog, click the "Hive Transformation Pipeline" template and click Add. Go back to the portal and copy the JSON from the TestPipeline and paste it into Visual Studio.

Once you save the pipeline, the pipeline screen will split into two with the diagram of the pipeline on the top and the JSON on the bottom, as shown in Figure 7-9.

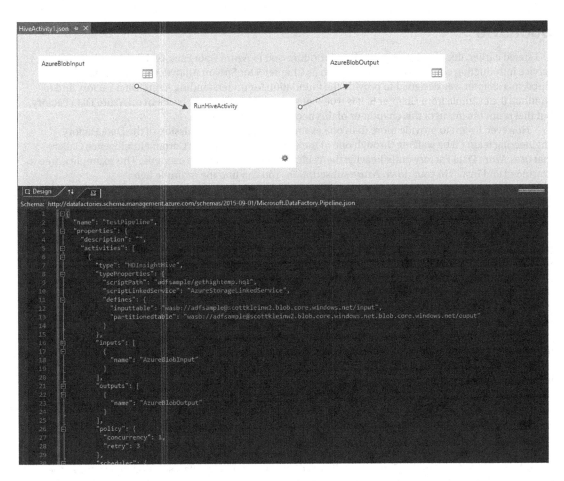

Figure 7-9. *Saved pipeline in Visual Studio*

Again, this was a simple example, but through it hopefully you saw how powerful yet easy it is to author a Data Factory pipeline. Publishing the Data Factory to Azure is as simple as right-clicking the project in Solution Explorer and clicking Publish. A Publish dialog will appear to walk you through publishing the Data Factory.

On the Configure Data Factory page, select the "Create New Data Factory" option, and enter a unique name for the Data Factory, select the appropriate subscription, resource group, and region, and then click Next.

On the Publish Items page, make sure all the Data Factory objects are selected, and then click Next. On the Summary page, click Next to begin the deployment process. Visual Studio uses the underlying REST APIs that were installed as part of the installation earlier of the Azure Data Factory Tools for Visual Studio to deploy the Data Factory pipeline. Once the deployment is done (it shouldn't take too long), you can go back into the portal to refresh your list of Data Factories, and on the Data Factory blade for your named Data Factory, click the Diagram tile. From there you will be able to monitor and troubleshoot the Data Factory normally.

Scenario

As I stated earlier, this chapter was simply to introduce you to Azure Data Factory. Like the other Azure service introduction chapters in this book, such as Chapter 5 for Stream Analytics and Chapter 3 for IoT Hubs, this chapter was designed to provide the foundation for understanding Azure Data Factory and to continue the example from Chapter 6. It is true that several chapters could be spent on Azure Data Factory, but that is not the intent of this chapter or of this book.

However, I want to provide more than one example, and so with permission of the Data Factory engineering team I'll be walking through one of the examples found in the Cortana Intelligence Gallery that uses Azure Data Factory quite heavily: the Vehicle Telemetry Analytics example. This example is free to download and install in your down Azure subscription. You can find the example here:

`https://gallery.cortanaintelligence.com/SolutionTemplate/Vehicle-Telemetry-Analytics-9`

On the top right of this page is a link to deploy this solution to your own subscription, which I highly recommend. Toward the bottom of the page is a link to the playbook and technical guide, which I also highly recommend you download. This guide walks you through an architecture overview of the solution, the services involved in the solution, and includes a tab in which you can deep dive into the solution.

The foundation of this example is based simply on the fact that the technology in the cars you and I drive has improved drastically. Cars today contain countless sensors that track nearly every aspect of the vehicle, and many of these sensors are connected to the Internet. Yes, connected cars. As stated in the link, it is predicted that by 2020, most cars on the road will be connected to the Internet.

So, given where the connected car topic is going, it is not hard to imagine how car manufacturers can use information to improve the safety and reliability of our cars. Hey, if it improves my driving experience, I'm all for it. On the flip side, auto insurance companies can use this information to determine patterns in my driving habits. Personally, I'm not keen on this. The last thing I need is for my auto insurance company to give me a call. "Mr. Klein, we noticed you were quite heavy on the gas and hard on the brakes today." I'd be yelling "GET OUTTA MY CAR!" into the phone. It's too big brother-ish. However, if I'm a good driver and my great driving habits were noticed by the auto insurance company and they used this information to lower my rates due to my stellar driving record and habits, ca-ching!

But you see where this is going, and thus this example, which demonstrates how Azure Data Factory and other Azure services of the Cortana Intelligence suite are used to create a solution that captures the essence of gaining real-time insights and predictive analysis on driving habits and vehicle performance.

The solution uses a number of Azure services from the Cortana Intelligence suite, including

- **Event Hubs**: Used for vehicle telemetry data ingestion.

- **Stream Analytics**: Following both hot and cold paths, Stream Analytics is used for real-time insights via Power BI dashboards and long term storage for downstream batch analysis.

- **Machine Learning**: Used to detect anomalies via predictive analysis.

- **HDInsight**: Used for large-scale data processing.

- **Data Factory**: Responsible for the orchestration, scheduling, management, and monitoring of the pipeline.

Using these services together results in the architecture shown in Figure 7-10.

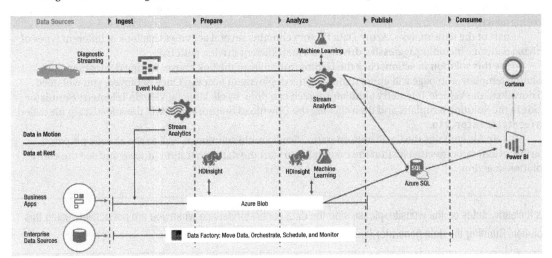

Figure 7-10. *Vehicle telemetry solution diagram*

Given the above architecture, the Azure Data Factory actually has a significant part in the entire process and solution. An application called the vehicle telematics simulator produces vehicle data, including diagnostic information, vehicle driving patterns, and sensor information. This data is then sent to Azure Event Hubs.

An Azure Stream Analytics job picks up data from both Event Hubs and Azure Blob Storage, and three different queries in the ASA job send the data to three different outputs: Azure SQL Database, Azure Blob Storage, and another Event Hub. The Azure Blob Storage data that the ASA job picks up is simply vehicle catalog reference data containing VIN-to-car model mapping.

Azure Data Factory is then responsible for managing the coordination of many of the other processes, as seen in Figure 7-11. These processes include generating additional simulated vehicle data for richer batch analytics and to ensure a decent data volume representation.

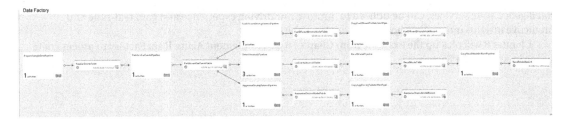

Figure 7-11. *Vehicle telemetry orchestration pipeline*

Moving further down the diagram, to the right you can see how other pipelines are used to operationalize the transformation and processing of batch data by kicking off HDInsight and Azure Machine Learning operations. The Azure Machine Learning operation looks for anomalies in the data and sends any anomaly data to Azure Blog storage.

As part of the pipeline workflow, the batch scoring endpoint of an Azure Machine Learning web service endpoint is registered as a Data Factory linked service and then operationalized using a Data Factory batch scoring activity.

As part of the data analysis, Azure Data Factory contains several activities that look at different types of driving patterns, including aggressive driving and fuel efficient driving patterns.

To see this solution in action, click the Deploy button from the link above. During deployment, a solution template web page will appear, showing the deployment process. Once deployed, you will need to download the Vehicle Telemetry Simulator, which can done by clicking the Vehicle Telemetry Simulator node in the Solution template, and then clicking the Download button, which will download a zip file called careentgenerator.zip.

This zip file contains the CarEventGenerator Visual Studio solution as well as the already prepared CarEventGenerator executable. Run the executable to start the data generation process and see the entire solution in action.

As it clearly states on the website, please stop the data generator service when you are not actively using this solution. Running the data generator (CarEventGenerator.exe) incurs costs.

Your homework assignment is to review the Azure Data Factory for this solution, including the pipelines, activities, datasets, and linked services, to see how this solution is constructed. However, remember to shut down the CarEventGenerator executable when you are not working with this solution.

Summary

This chapter provided an overview of Azure Data Factory, beginning with a discussion about the need for a platform in which existing services can collect, aggregate, and process data from disparate systems and produce useful information. This chapter discussed the key benefits, characteristics, and components of Azure Data Factory, discussed where Azure Data Factory fits into the Azure data services ecosystem, and provided examples of how Azure Data Factory can be leveraged in the process to gain valuable data insight.

This chapter then walked through creating a simple Data Factory using both the Azure portal as well as Visual Studio to help build the foundation for understanding how reliable, repeatable, and deterministic data movement and transformation pipeline orchestrations are built and utilized.

Lastly, this chapter provided an example of how Azure Data Factory, along with other Azure Cortana Intelligence suite services, can be utilized to improve our driving experience and gain real-time and predictive insights into vehicle health and driving habits.

Chapter 8 will pick up the example from Chapter 6 and implement Azure Data Factory to process data sitting in Azure Data Factory.

■ ■ ■

Integrating Data Between Data Stores Using Azure Data Factory

Chapter 7 introduced Azure Data Factory, its collaboration benefits and capabilities for data collection, aggregation, and processing. Chapter 7 also walked through the creation of a sample Data Factory pipeline using both the Azure portal as well as Visual Studio, and then closed out the chapter by taking a look at the Vehicle Telemetry Analytics example, a real-world scenario showcasing how Azure Data Factory can be used to orchestrate data movement and integration. With that information as the foundation, it is time to continue the example from Chapter 6 and implement Azure Data Factory.

So far in your example, data is sitting in both Azure Data Lake Store and Azure Blob Storage, waiting for the next step in the journey. This chapter will apply what was learned in Chapter 7 and build an Azure Data Factory pipeline to integrate other data and move data into other storage services.

Building the Pipeline

The first step in creating the pipeline is to create the Data Factory and prepare the environment. For the Data Factory, this example will pick up the data from Azure Blob Storage and use a stored procedure activity to move the data to Azure SQL Database. You'll also use a Data Factory to pull external data into the pipeline for reference data.

Preparing the Environment

The first step is to create the Data Factory, so in the Azure portal (`portal.azure.com`) click New and then select Data + Analytics. Then scroll down in the Data + Analytics blade and click the Data Factory option.

In the New Data Factory blade, enter a name for the Data Factory and select the appropriate Azure subscription. For the Resource Group, select "Use existing option," and then select the resource group that you have been using throughout this example. Lastly, select the location (region) that you have put all the other resources in. Your New Data Factory blade should look similar to Figure 8-1. To make things easier, also check the "Pin to dashboard" checkbox and then click Create.

© Scott Klein 2017
S. Klein, *IoT Solutions in Microsoft's Azure IoT Suite*, DOI 10.1007/978-1-4842-2143-3_8

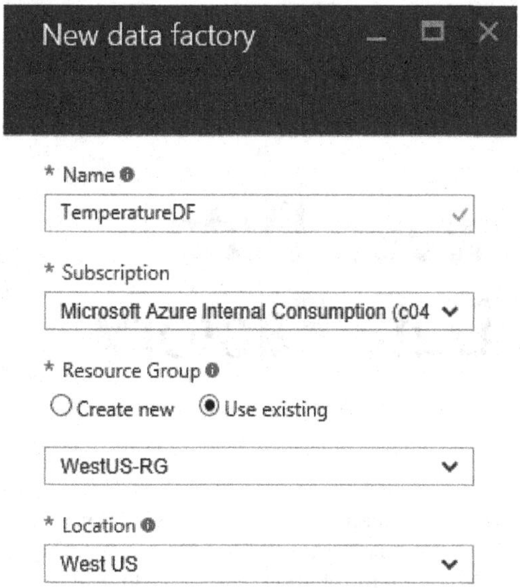

Figure 8-1. *Creating an Azure Data Factory*

The creation of the Data Factory only takes a few seconds, after which you will be presented with the similar Properties and Settings blades. Before you start building the pipeline, there is a bit more environment preparation that needs to take place. Since the pipeline will be sending data to Azure SQL Database, the table in which the data will be stored along with the stored procedure that will be copying the data both need to be created.

If you have an existing Azure SQL Database server and database, feel free to use that one, or create a new one. As a quick reference, I'll briefly walk through creating the server and database in which the table and stored procedure will be created.

In the Azure portal, click New, select Data + Storage, and then select SQL Database. This will open up the SQL Database blade. In the blade, enter a new database name and select the same subscription you have been using. For the Resource Group, select "Use existing option," select the resource group that you have been using throughout this example, and then select "Blank database" for the source.

For the server, select the server from the list that matches the region in which your resource group resides, or create a new server by clicking "Create a new server." If you create a new server, ensure that the location is in the appropriate region, make sure the Create V12 server is set to Yes, and ensure that the "Allow Azure Services to access server" option is checked. Remember the username and password you enter because you will need that shortly.

Back on the SQL Database blade, select the Pricing Tier option. There are quite a few options and it is beyond the scope of this book to discuss all of them. There are plenty of resources that discuss pricing and performance, and a good place to start is here:

https://azure.microsoft.com/en-us/pricing/details/sql-database/

However, for the purposes of this example, scroll down in the Choose Your Pricing Tier blade, select Basic, and then click Select. Once the SQL Database blade is filled out, click Create. It will take a few moments to provision the database. Once the database is created, the Properties blade for the database will appear.

In the Properties blade, click the Server name link, which will open the Settings and Properties blades for the server you either selected or created. In the Properties blade, click the "Show firewall settings" link, which will open the Firewall settings blade. This blade shows the IP address from which you are currently connected. In order to connect to the SQL database you created, you will need to add a firewall rule that allows you to connect from the current IP address. (See Figure 8-2.)

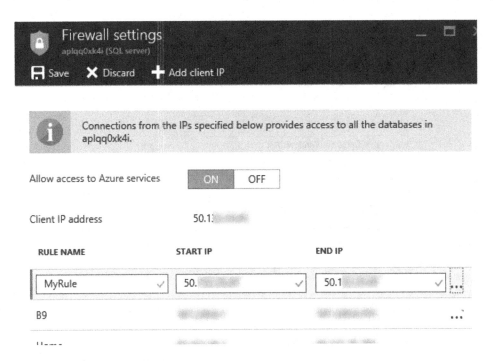

Figure 8-2. *Setting the firewall rules for the server*

If you have an existing firewall rule, you don't need to add it again. If you have not added the rule, enter a rule name, copy the IP address into the Start IP and End IP boxes, and then click Save.

The "Allow access to Azure services" option allows applications from Azure to connect to the selected Azure SQL Server. When an attempt is made to connect to data on this server from an application from Azure, the firewall verifies that Azure connections are allowed. Setting this option to On allows Azure connections, while Off means that requests do not reach the Azure SQL Database server.

With the server and database created and configured, the next step is to create the table and stored procedure objects. You can use either SQL Server Management Studio or Visual Studio (and SQL Server Data Tools) for this. You can download SQL Server Management Studio here:

`https://msdn.microsoft.com/en-us/library/mt238290.aspx`

When launching SQL Server Management Studio, the Connect to Server dialog will appear, as shown in Figure 8-3. In this dialog, enter the complete server name and the username and password you used when creating the server. The server name can be obtained from the SQL Database Properties blade (you clicked it a few minutes ago to get to the server firewall rules).

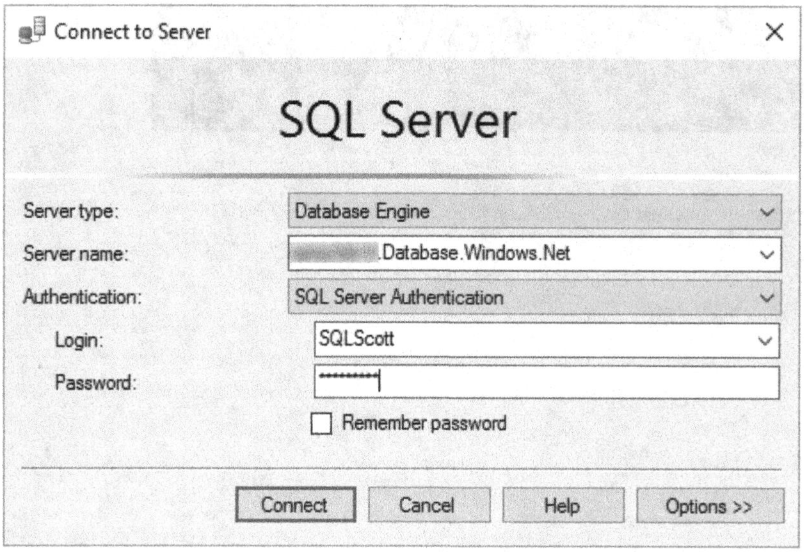

Figure 8-3. *Connecting to the server*

In the early days of Azure SQL Database you used to have to use the *username@servername* format as the user login to log in (for example, *SQLScott@servername*). This is no longer the case, although you will still find a lot of examples online that specify the *username@servername*. However, today you only need to specify the login name.

Click Connect, at which point SQL Server Management Studio will load with the Object Explorer pane on the left. Expand the Databases node, click the database you just created in the portal, and then click the New Query button on the toolbar. Enter the following code into the query window. Before executing the code, ensure that the database you clicked is actually displayed in the drop-down on the toolbar next to the Execute button. This will save you from adding the table in the wrong database. If it all looks good, click the Execute button.

```
/* Table Creation */
SET ANSI_NULLS ON
GO

SET QUOTED_IDENTIFIER ON
GO

CREATE TABLE [dbo].[DeviceTemps](
    [ID] [int] IDENTITY(1,1) NOT NULL,
    [Device] [nvarchar](50) NOT NULL,
    [Sensor] [nvarchar](50) NOT NULL,
    [Temp] [nvarchar](50) NOT NULL,
    [Humidity] [nvarchar](50) NOT NULL,
    [GenDate] [datetime] NOT NULL,
    [InsertDateTime] [datetime] NULL,
 CONSTRAINT [PK_DeviceTemps] PRIMARY KEY CLUSTERED
(
    [ID] ASC
```

```
)WITH (PAD_INDEX = OFF, STATISTICS_NORECOMPUTE = OFF, IGNORE_DUP_KEY = OFF, ALLOW_ROW_LOCKS
= ON, ALLOW_PAGE_LOCKS = ON)
)

GO

ALTER TABLE [dbo].[DeviceTemps] ADD  CONSTRAINT [DF_DeviceTemps_InsertDateTime]  DEFAULT
(getdate()) FOR [InsertDateTime]
GO
```

This was a bit of work to get everything ready, but it was necessary, especially if you don't have experience with Azure SQL Database (I can't make any assumptions). At this point, you can continue creating the pipeline, focusing now on creating the linked services.

Creating the Linked Service

As you learned in the previous chapter, the next step in building and creating the Data Factory pipeline is creating the appropriate linked services. Back in the portal on the dashboard, click the Azure Data Factory tile for the Data Factory you created above. There are two linked services that need to be created, which will define the connections to the Azure Blob Storage and Azure SQL Database just created. Let's begin with the Azure Storage data store, so in the Properties blade, click the Author and Deploy tile.

Azure Storage

Similarly to the steps you learned in the previous chapter, click the "New data store" link and then select Azure Storage from the list. In the Draft blade, enter the following code and replace the AccountName and AccountKey with the appropriate values for the Azure Storage account that the temperature sensor data is sitting in:

```
{
    "name": "AzureStorageLinkedService",
    "properties": {
        "description": "",
        "type": "AzureStorage",
        "typeProperties": {
            "connectionString":
"DefaultEndpointsProtocol=https;AccountName=<AccountName>;AccountKey=*********"
        }
    }
}
```

Be sure to remove any brackets (< or >) from the AccountName and AccountKey, and then click Deploy.

Azure SQL

The next step is to create the linked service to the Azure SQL Database created above. Again, click the "New data store" link and select Azure SQL Database. In the Draft blade, enter the following code and replace the <username>, <servername>, <databasename>, and <password> with the appropriate values:

```
{
    "name": "AzureSqlLinkedService",
    "properties": {
        "description": "",
        "hubName": "temperaturedf_hub",
        "type": "AzureSqlDatabase",
        "typeProperties": {
            "connectionString": "Data Source=tcp:<servername>.database.windows.
net,1433;Initial Catalog=TemperatureData;Integrated Security=False;User ID=<username>@<serve
rname>;Password=**********;Connect Timeout=30;Encrypt=True"
        }
    }
}
```

Click Deploy when finished. The great thing about deploying is that the deployment process actually validates the connection information, allowing you to correct any information long before you build and run the pipeline. This helps with troubleshooting.

Creating the Datasets

Next is the creation of the datasets. As you learned in the last chapter, datasets are simply references to the data that will be used. For example, the Azure SQL Database linked service points to the server and database the activity will connect to. The database specifies the table in which the pipeline activity will insert the data into.

The same goes for Azure Storage. The linked service simply defines the account the pipeline activity will connect to. The dataset specifies the container, folder, and file in which to get the data.

You'll begin with the Azure SQL dataset. In the portal, click the More link and select New dataset, and then select Azure SQL. In the Drafts blade, enter the following code:

```
{
    "name": "AzureSQLDataset",
    "properties": {
        "type": "AzureSqlTable",
        "linkedServiceName": "AzureSqlLinkedService",
        "structure": [],
        "typeProperties": {
            "tableName": "DeviceTemps"
        },
        "availability": {
            "frequency": "Week",
            "interval": 1
        }
    }
}
```

The linkedServiceName property specifies the Azure SQL linked service created above, and the tableName specifies the Azure SQL Database table that was created earlier in this chapter. The frequency and interval are set as monthly, which specifies the availability of the input slices, in this case monthly. Go ahead and click Deploy to save and deploy this dataset.

Next you'll create the Azure blob dataset. In the portal, click the More link, select New dataset, and then select Azure Blob Storage. In the Drafts blade, enter the following code:

```
{
    "name": "AzureBlobInput",
    "properties": {
        "type": "AzureBlob",
        "linkedServiceName": "AzureStorageLinkedService",
        "typeProperties": {
            "fileName": "data.csv",
            "folderPath": "adfsample/input",
            "format": {
                "type": "TextFormat",
                "columnDelimiter": ","
            }
        },
        "availability": {
            "frequency": "Week",
            "interval": 1
        },
        "external": true,
        "policy": {}
    }
}
```

The linkedServiceName property specifies the Azure Storage linked service created above, the filename specifies the name of the file in Azure Storage in which you want the pipeline activity to pick up its data, and the folderpath property specifies the container and folder in which the filename resides. The availability section defines the processing window, such as how often the data should be processed to give a slicing model for the dataset. Options include a frequency of minute, hour, day, week, and month. In this example, the frequency is set to every week. The external property specifies whether or not the dataset is explicitly produced by a data factory. Unless a dataset is being produced by Azure Data Factory, this property should be marked as external, or *true*. Click Deploy.

Creating the Pipeline

Lastly, it is time to create the actual pipeline. In the portal, click the More link and select "New pipeline." The Drafts blade will populate with the template of a pipeline, so you'll need to add the appropriate activity. In the Drafts blade, click the "Add activity" link, and select "Copy activity."

Fill in the pipeline and activity as shown in the following code:

```
{
    "name": "TemperaturePipeline",
    "properties": {
        "description": "Copy data from blob storage to an Azure SQL table",
        "activities": [
            {
                "type": "Copy",
                "typeProperties": {
                    "source": {
                        "type": "BlobSource"
                    },
                    "sink": {
                        "type": "SqlSink",
```

```
                        "writeBatchSize": 10000,
                        "writeBatchTimeout": "60.00:00:00"
                    }
                },
                "inputs": [
                    {
                        "name": "AzureBlobDataset"
                    }
                ],
                "outputs": [
                    {
                        "name": "AzureSQLDataset"
                    }
                ],
                "policy": {
                    "timeout": "01:00:00",
                    "concurrency": 1,
                    "executionPriorityOrder": "NewestFirst",
                    "retry": 2
                },
                "scheduler": {
                    "frequency": "Week",
                    "interval": 1
                },
                "name": "TemperatureCopyActivity",
                "description": ""
            }
        ],
        "start": "2016-08-01T00:00:00Z",
        "end": "2016-08-30T00:00:00Z",
        "isPaused": false,
        "hubName": "temperaturedf_hub",
        "pipelineMode": "Scheduled"
    }
}
```

Once it is all filled in, click Deploy. At this point, you could go see the pipeline run, but before you do that, let's see how you did building your pipeline by comparing it to one that is built using the Copy Data wizard.

Copying Data

You are probably asking why you went through the process of manually building the pipeline (and associated linked services and datasets) when you could have done this manually. It is a great question, and the answer that I will give is that it helps you understand how pipelines and activities are constructed.

The only wizard that exists today is the Copy Data wizard, probably because that is the activity that most people will use. Whether there will be a Sproc Activity wizard or another activity wizard in the future is unknown. And since the copy activity is the only one so far, it is helpful to understand how to construct the pipelines and activities manually.

Thus, back in the Data Factory blade, click the Copy Data tile, shown in Figure 8-4, to launch the Copy wizard.

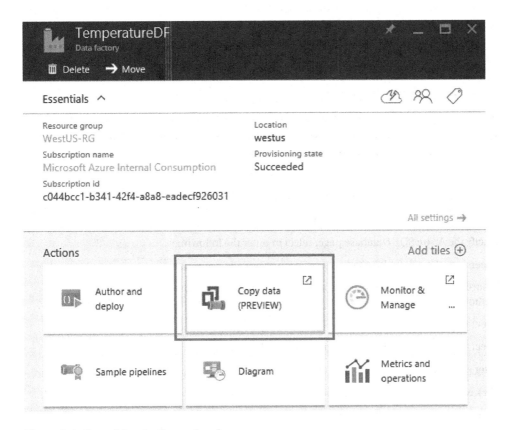

Figure 8-4. *Launching the Copy wizard*

On the properties page, enter the following:

- Task name and description

- Select "Run once now" or "Run regularly on schedule." If selecting "Run regularly on schedule," then provide a recurring pattern. Note the start date time and end date time, and enter the appropriate dates. Then click Next.

On the Source data store page, do the following:

- Ensure that "Connect to a Data Store" is selected, select the Azure Blob Storage icon, and then click Next.

On the Specify the Azure Blob storage account page, select the following:

- Either accept the default Linked service name or provide a new name.

- Ensure that "From Azure Subscription" is selected for the Account selection method.

- Select the appropriate Azure storage service where the temperature data is stored, and then click Next.

On the Choose the input file or folder page, do the following:

- Select the appropriate folder (in this case, rawdata), click Choose, and then click Next.

131

On the File format settings page, select the following:

- Select the Text file format.

- Leave the Column delimiter as Comma.

- Leave the Row delimiter as Line Feed.

- Set the Skip header line count to 1.

- Notice at the bottom of the page the wizard populates a grid with sample data from the file found in the rawdata folder.

- Click Next.

On the Destination data store page, do the following:

- Select the Azure SQL Database icon and then click Next.

On the Specify the Azure SQL Database page, select or enter the following:

- Either accept the default linked service name or provide a new name.

- Ensure that From Azure Subscription is selected for the server/database selection method.

- Select the appropriate server name from the drop-down.

- Select the appropriate database name from the drop-down.

- Enter the user name and password of the server.

- Click Next.

On the Table mapping page, do the following:

- Select the DeviceTemps table from the drop-down.

- Click Next.

On the Schema mapping page, do the following:

- Ensure that the mappings are correct. For example, Column0 should map to the Device column, Column1 should map to the Sensor column, and so on.

- Leave the Repeatability settings method to None.

- Click Next.

On the Summary page, ensure that all of the information is correct and then click Finish. After a few moments, the entire data factory pipeline will be created, including the linked services, datasets, and pipeline using the Copy activity.

Once it is created, close the wizard page, and go back to the Azure portal. In the Data Factory Author and Deploy blade you will see the new linked services, datasets, and pipeline created by the wizard, as shown in Figure 8-5.

Figure 8-5. *The wizard took care of everything*

Click the new pipeline that was created by the Copy wizard to open it up in the Draft pane. You can see that what you coded above is actually pretty close to what the wizard created. You'll notice a few additional items.

First, in the `BlobSource` property, you should notice the `skipHeaderLineCount` property. This was selected when creating the linked service in the wizard. This tells the copy activity to skip the first line of the file, which contains the column names, not actual data. The pipeline will fail if this is not set.

Also notice the `recursive` property, which is set to false. This property simply tells the copy activity to also process all files in any subfolders of the `rawdata` folder. This could be set to Yes but since the `rawdata` folder has no subfolders, you can leave this as false.

The `SqlSink` has different `writeBatchSize` and `writeBatchTimeout` values. This isn't critical but it is good to understand.

Also notice the addition of the translator type section, with a type and `columnMappings` property.

Other than that, it is pretty much the same and good to go. Let's go see how the pipeline is doing.

Running and Monitoring the Pipeline

Back on the Data Factory blade, click the Monitor and Manage tile (it is next to the Copy data icon), which will open the Data Factory Monitoring page and show all the pipelines you have created so far pertaining to the Data Factory in which the pipeline was created.

The Monitoring page is broken up into several sections. On the left is the Resource Explorer that looks very similar to the Data Factory blade in the portal for creating new linked services, datasets, and pipelines. The center section is split into two areas: the top shows the actual pipelines with their associated datasets for the selected Data Factory, and the bottom area shows activity for the all the pipelines in the selected Data Factory. The right pane is the Properties pane, which shows properties for any items selected in the top middle section.

In the top section, you can pause or stop the pipelines, and in the lower Activity section you can double-click any activity to drill into the status of that activity, whether it is pending, waiting, or failed. If a pipeline failed, drilling into the activity will allow you to see why and where it failed.

To verify that the pipeline indeed worked, open up SQL Server Management Studio and query the `DeviceTemps` table. All the rows from the text file in Azure Blob storage should be copied to the `DeviceTemps` table, as shown in Figure 8-6.

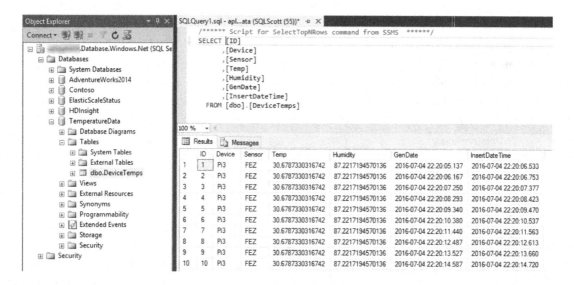

Figure 8-6. *All of the rows were copied*

Congratulations, you now have a working Data Factory pipeline. Before we close out this chapter, there is one other activity that should be discussed.

Monitoring and Managing Azure Data Factory

The previous chapter and this chapter spent quite a bit of time looking at the components and aspects of Azure Data Factory in order to move and transform data between various data stores. As you have learned, there are many areas and moving parts of Azure Data Factory, and in a multifaceted system such as Azure Data Factory, gaining insight into what is happening in the service is critical to ensure the service is running smoothly. Azure Data Factory does just that by providing an interface to help troubleshoot and debug Data Factory issues as well as monitor and manage Data Factory pipelines.

The Monitoring and Management application of Azure Data Factory delivers the insight needed to easily troubleshoot and manage the many aspects of Azure Data Factory. To open the Monitoring and Management application, click the Monitor and Manage tile in the Azure Data Factory overview blade, show in Figure 8-7.

You might be wondering what the difference is between the Monitor and Manage tile and the Metrics and Operations tile. Azure Data Factory allows you to capture different metrics and create alerts on those metrics. The Metrics and Operations tile allows you to view those metrics for historical purposes, whereas the Monitor and Manage tile provides an avenue for troubleshooting user events.

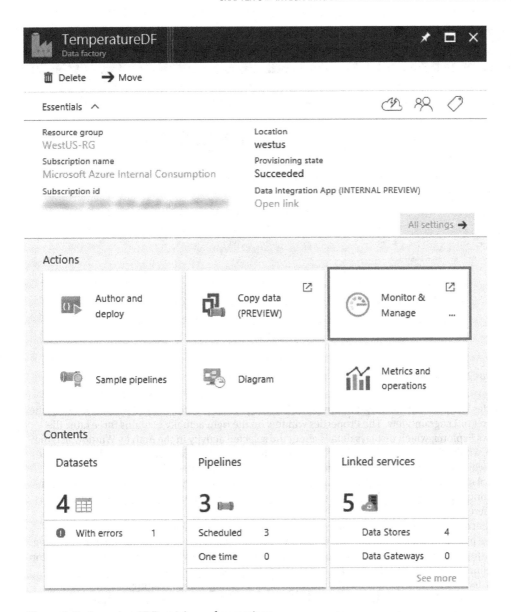

Figure 8-7. Accessing ADF metrics and operations

Clicking the Monitor and Manage tile opens the Monitor and Management application in a new browser window, shown in Figure 8-8. The Resource Explorer pane on the left shows all the resources in the Data Factory, including datasets, linked services, and pipelines. The Diagram View on the top and middle provides a visual, end-to-end view of the Data Factory pipelines and associated assets to help manage, monitor, and troubleshoot issues.

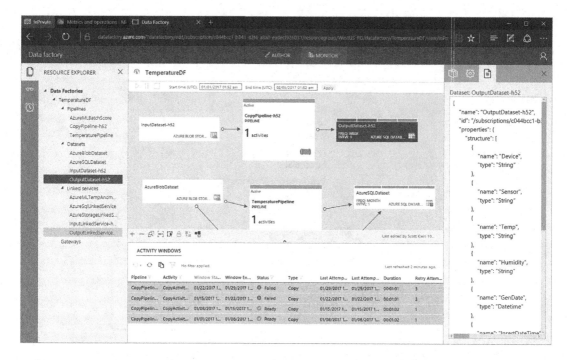

Figure 8-8. *Azure Data Factory Monitor*

The Activities Window in the center, bottom pane lists all activities for the selected datasets in either the Resource Explorer or Diagram View. The Properties window on the right actually contains three tabs: the Activity Windows Explorer, which displays details about the selected activity in the Activity Window list; the Properties page, which displays properties for the time selected in either of the first three windows; and the Script window, which shows the JSON definition of the selected Data Factory entity.

There is a lot of information packed into that single browser window and on the surface it could seem a bit overwhelming, but as you start clicking around you'll quickly realize the relationships and interactions between the different panes and windows, and how easy it is.

Pipelines that are enabled and not in a paused state are shown with a green line. Selecting a pipeline enables the three play, pause, and stop buttons to allow you to interactively work with the pipeline. Pressing the stop button stops the pipeline and terminates any existing executing activities. Pressing the pause button pauses the pipeline but allows any currently running activities proceed to completion.

Activities listed in the Activity Windows list are listed in descending order, meaning that the latest activity is at the top of the window. You should know that the Activity Window list does not refresh automatically; you need to click the refresh button to refresh the list. You can filter the list of activities in the list by changing the start time and end time values above the Diagram View. Double-clicking a specific activity in the Activity Window opens the Activity Window Explorer on the right, displaying detailed information about the selected activity, including failed attempts and more.

Additional information is available by clicking in the bar over the dataset, as shown in Figure 8-9. In this figure, the pipeline has failed writing to the target dataset, and clicking the top bar of the dataset displays an activity window sliced by a frequency of one week intervals. You can then click any of the individual time slices in the pop-up activity window to have the details of that timeslice display in the Activity Window Explorer on the right.

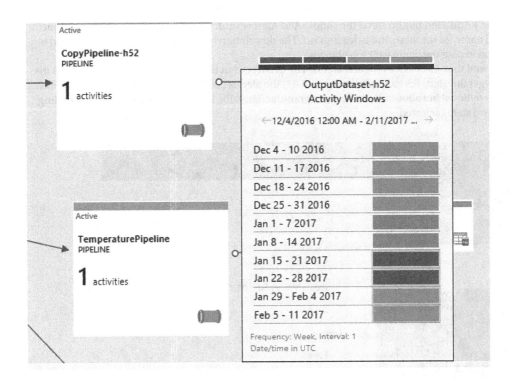

Figure 8-9. Activity Window

Alerts

Alerts are a fantastic way to proactively know what is happening during a pipeline execution, which makes troubleshooting and monitoring Azure Data Factory much easier. Creating an alert is as simple as clicking the Alert icon on the left pane and then clicking the + Add Alert button, shown in Figure 8-10.

The Alerts pane will also show any existing alerts and allow you to edit, delete, or disable/enable them.

Figure 8-10. Adding an alert

Clicking the + Add Alert button starts the simple Add Alert wizard. On the first page, provide a unique and meaningful name for the alert, and a description. The description is optional but it is recommended to add a description in case the name isn't too obvious.

Click Next, and then select an event and status. The event is what triggers the alert, and the status is the condition to trigger the alert. For example, in Figure 8-11, the alert is configured to trigger if an activity fails. The substatus is optional but allows you to be more granular. The substatus options will change depending on the activity and status selected.

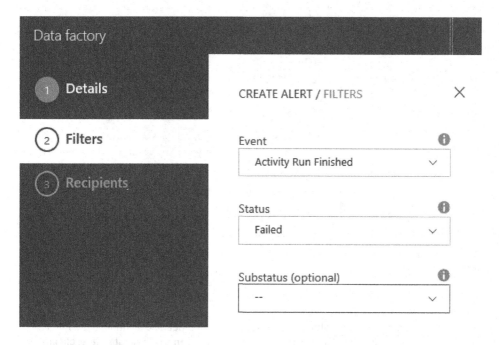

Figure 8-11. *Configuring the alert*

Click Next, and then enter the email addresses of the recipients you want the alert to notify. Multiple email addresses can be entered by separating them with a comma. Click Finish to save the alert. The alert will then be listed on the main Alerts page where you can then add additional alerts or edit, delete, or enable/disable them.

OK, this was a quick overview of the Monitor and Manage application for Azure Data Factory. Again, I highly recommend that you spend time clicking around in here to better familiarize yourself with its capabilities. As you work with Azure Data Factory, this page will serve you well.

Summary

This chapter continued the example from Chapter 6, which picked up the data streamed into Azure Storage by Azure Stream Analytics. In this chapter, you created a Data Factory, and then built an orchestration pipeline which consisted of a couple of linked services and datasets, and a pipeline containing a data copy activity to copy the data from Azure Storage to Azure SQL Database.

You then used the Copy data wizard, which walks through the creation of a copy data pipeline and associated linked services and datasets, and you then compared the two pipelines. You did this in order to understand the properties necessary to define the pipeline.

Lastly, you launched the Monitoring page, which provides the ability to view the pipeline activities, and troubleshoot and correct any issues with the pipeline.

Next, you will look at the analytical and processing services, beginning with Azure Data Lake Store.

PART III

Data at Rest

CHAPTER 9

■ ■ ■

Azure Data Lake Store

The chapters in the last section focused completely on data on the move, that is, data that is being generated by devices and routed through communication, streaming, and transformation services, ultimately ending up in a storage mechanism for further analysis and processing. Chapter 6 routed the incoming data to two different data stores, Azure Blob Storage and Azure Data Lake Store, using Azure Stream Analytics.

As you saw in Chapter 6, other outputs were available to store the data including Azure DocumentDB and Azure SQL Database. DocumentDB is an interesting one because since your data is coming in as JSON, your data could have easily used DocumentDB as the data store. DocumentDB is a fully managed NoSQL JSON database designed for modern applications, including IoT. Thus, DocumentDB could certainly have been an effective choice for storing the incoming data.

So why were Azure Data Lake Store and Azure Blob Store chosen as the outputs for the Stream job? For two reasons, honestly. Azure Blob Storage was included to show that multiple outputs can be included in a query and the power that offers when routing data. One query can route certain data to one output and another query can route other data to another output. As you saw in the Azure Stream Analytics chapter, a single Azure Stream Analytics job used two different queries to route data to two different outputs, each with a different format. Very powerful.

So why Azure Data Lake Store? The answer to that question is found in the example used throughout this book. This example is taking data from several Raspberry Pis and a Tessel, and routing that data through IoT Hub and Stream Analytics. Quite easy, but as pointed out in an earlier chapter, imagine thousands or tens of thousands of these devices sending data every second or every minute. Imagine the workload flowing through these services and the amount of data being generated. Next, imagine needing to process this data quickly and efficiently for real-time or near real-time insights. The example in this book is using device temperature information, but imagine that this data is banking information and you're looking for fraud detection, another example of information you need to know about ASAP. Additionally, the data store should scale to meet the demands of the client to optimize performance.

While the data generated in your example could probably be stored in Azure Blob Storage, Azure Blob Storage is a general purpose storage mechanism, and it specializes in storing unstructured data as objects. What your scenario, or any other real-time analytic application scenario, needs is a hyperscale repository specifically designed for big data analytical workloads without any limitations on type or size. Enter Azure Data Lake Store.

Azure Data Lake

To be clear, there is no Microsoft product or cloud service called "Azure Data Lake." Now, to be completely honest, not long ago there used to be, but it evolved into what is now called Azure Data Lake Store. Today however, what one would call *Azure Data Lake* is actually three individual services for storing and analyzing big data workloads: Azure Data Lake Store (ADLS), Azure Data Lake Analytics (ADLA), and Azure HDInsight (HDI). So, when you say "Azure Data Lake," you are actually referring to a small collection of services aimed at making big data analytics easy.

© Scott Klein 2017
S. Klein, *IoT Solutions in Microsoft's Azure IoT Suite*, DOI 10.1007/978-1-4842-2143-3_9

This chapter will focus entirely on Azure Data Lake Store, with Chapter 10 discussing Azure Data Lake Analytics, Chapter 11 discussing the U-SQL language, and Chapter 12 covering Azure HDInsight. The following sections will discuss each of these services and technologies briefly to help demystify and provide clarity on each of these services before exploring each of them further in the upcoming chapters.

Azure Data Lake Store

Simply put, Azure Data Lake Store is a hyperscale data repository for the enterprise for big data analytic workloads. Unlike Azure Storage where there is a limit to file size, ingestion speed, and other limitations, Azure Data Lake Storage has no limitations regarding the size or types of files, nor the speed at which data is ingested or consumed.

Accessible through WebHDFS-compatible REST APIs, Azure Data Lake Store enables big data analytics tuned for performance and includes the needed benefits for enterprise scenarios such as security, scalability, and reliability. HDFS will be discussed more in Chapter 12, but I'll briefly give an overview here, as well as WebHDFS. HDFS, or Hadoop Distributed File System, is a distributed file system that is part of Apache Hadoop and provides high-throughput access to data. WebHDFS is an HTTP REST API that provides a FileSystem interface for HDFS.

Azure Data Lake Analytics

Azure Data Lake Analytics is all about writing, running, and managing data analytic jobs. Azure Data Lake Analytics removes the complexity of managing distributed infrastructure and lets you focus on the data and extracting the needed insights. ADLA dynamically provisions resources to meet the scaling needs of your analytic job, easily handling petabytes and exabytes of data. ADLA will be discussed in more detail in Chapter 10.

U-SQL

Included with Azure Data Lake Analytics is a query language called U-SQL, or Unified SQL, a language that combines the declarative and powerful query language of SQL with the expressive power of C#, allowing developers to build analytical jobs using their existing skills. U-SQL will be covered in Chapter 11.

Azure HDInsight

Azure HDInsight is a scalable process and analysis cloud service based on the Hortonworks Data Platform (HDP) Hadoop distribution. Deploying an Azure HDInsight cluster provisions a managed Apache Hadoop cluster, delivering a highly scalable framework for processing and analyzing big data workloads. Hadoop typically refers to the distributed components as a whole, including Apache HBase, Spark, Storm, and other technologies in the same ecosystem. However, for our purposes, Hadoop refers to the query workload and programming model to process and analyze data. More on Hadoop and HDInsight in Chapter 12.

With this brief introduction, the remainder of this chapter will focus on Azure Data Lake Store.

Azure Data Lake Store

Simply put, Azure Data Lake Store is a hyperscale data repository for the enterprise for big data analytic workloads. ADLS lets you store data of any size and type in a single place designed for analytics on the stored data.

Several times during this book the term "hyperscale" has been used, but what does "hyperscale" really mean? This term has reference to distributed computing and refers to a platform's ability to dynamically and efficiently scale from one or a few servers to many servers, such as hundreds or thousands depending on workload. Hyperscale computing is most often seen in cloud computing with big data type workloads, so it makes sense to see hyperscale available and used in ADL, ADLS, and HDinsight.

ADLS (and ADLA) was built with the enterprise capabilities you would expect in an analytical data store including security, reliability, and scalability. A key capability of ADLS is that it is built with hyperscale distributed file systems in mind, including Hadoop and the Hadoop Distributed File System. Because of this, its data can be accessed from Hadoop via an HDInsight cluster, and the data is accessible using WebHDFS-compatible APIs. This enables existing applications or services that use WebHDFS APIs to easily take advantage of ADLS.

Data within ADLS is also accessible through the Azure Portal, Azure Powershell, and SDKs for .NET, Node.js, and Java. All of these interfaces allow you to work with the data in ADLS, including creating folders and files, uploading and downloading data, and more.

It's worth mentioning again that ADLS does not enforce any limitations on the type and size of data. Files can range from kilobytes to petabytes or exabytes in size, and they are stored durably by making copies, thus making the data also highly available. The data is also highly secure by using Azure Active Directory for authentication and access control lists to manage who has access to the data.

The following section will walk through the creating of an Azure Data Lake Store.

Creating a Data Lake Store

The example in Chapter 6 showed how to create an Azure Data Lake Store. The simple process will be shown again here as a lead-in to working with the Azure Data Lake Store. Open the Azure Portal by navigating to Portal.Azure.com in your favorite browser. Once in the new portal, select New ➤ Intelligence + Analytics ➤ Data Lake Store.

The New Data Lake Store blade opens up and, as shown in Figure 9-1, there isn't a whole lot to enter or select to create a Data Lake Store. Simply give it a name, select the appropriate resource group and location, and click Create.

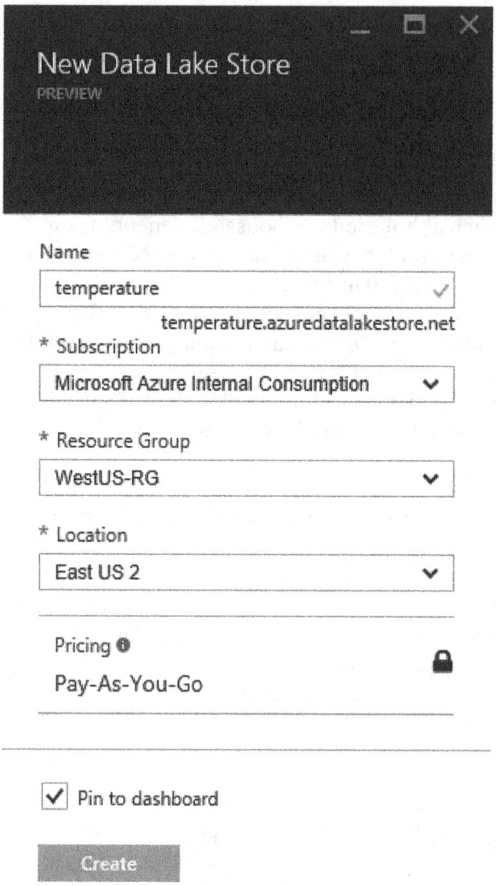

Figure 9-1. *Creating a New Azure Data Lake Store account*

Now, since this service is in preview as of this writing, the only location available to create an Azure Data Lake Store is the East US 2 data center. Many more data centers and regions will be available when this service GAs (is generally available).

In less than a minute the Azure Data Lake Store should be created and will display the Essentials and Settings blade shown in Figure 9-2.

I won't be spending a lot of time on Settings blade, but I will highlight a few things and discuss them later in the chapter. First is the Access Control (IAM) list. This area is where you define who has access to the Data Lake Store and what type of access they have.

The Firewall blade allows you to control access to the data in the Data Lake Store at the network level. By enabling the firewall and adding IP address ranges, only clients within that IP address range can connect to the Data Lake Store. This is a type of network isolation in which you can control access to your data at the network level.

The Data Explorer blade provides data management of the data in the Azure Data Lake, including creating and renaming folders, uploading new data, or defining permissions via the access control list.

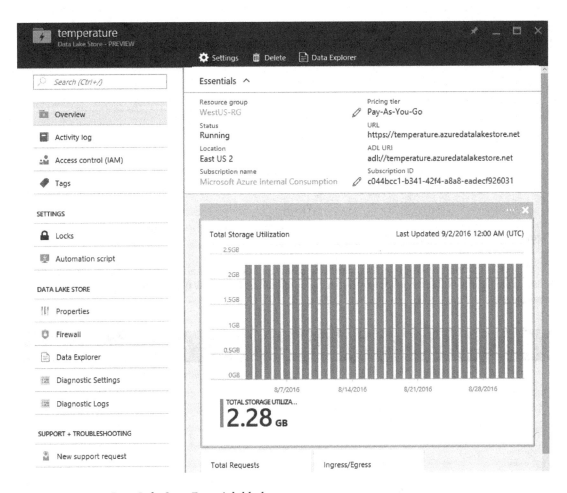

Figure 9-2. Azure Data Lake Store Essentials blade

At this point, let's turn our attention to actually working with Azure Data Lake Store. A little bit of time will be spent in the Azure Portal itself, and the rest of the time will be spent building an application to work directly with Azure Data Lake Store.

Working with Azure Data Lake Store

As stated above, there are a number of ways to work with Azure Data Lake Store, including Powershell and some APIs. The following two sections will show you how to use both the Azure Portal and the .NET SDK to perform a handful of operations against the Data Lake Store.

Azure Portal

Where this section will spend its time is in the Data Explorer. When you first open the Data Lake Store in the Azure portal you see the Settings blade, which allows you to monitor and manage the Data Lake Store, as shown in Figure 9-2. However, one of the critical links on the blade is the Data Explorer blade, as seen in Figure 9-3.

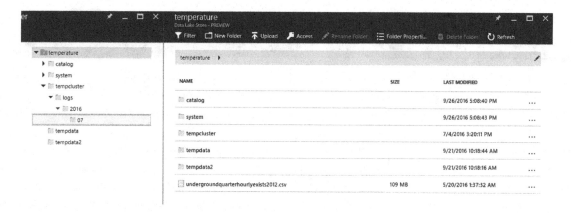

Figure 9-3. *ADLS Data Explorer blade*

The Data Explorer blade provides the means to navigate the folder and file structure of the Data Lake Store, as well as the ability to create new folders or subfolders, create new files, and define folder- and file-level access.

Figure 9-4. *ADLS folder and file structure*

Selecting a file opens the File Preview blade showing, by default, the first 25 rows of data in the file. Within the File Preview blade is also the option to download, rename, or delete the file, and the ability to look at the file properties, and specify when a file is automatically deleted via the Set Expiry option. The Set Expiry option is much like a garbage collection process in which you can clean up old data that isn't needed any longer. Simply specify the date and time of the file expiration, and when that date and time is reached for the particular file, the file will automatically be deleted by the ADLS service.

Notice in Figures 9-3 and 9-4 the folder structure from the output of the Azure Stream Analytics output into Azure Data Lake Store. In the output for the Data Lake Store the example defines the Path prefix pattern as tempcluster/logs/{date}, which results in the data being segregated by year and month.

However, it should be pointed out that unlike Azure Blog Storage, with Azure Data Lake store there is no limit to file size (as mentioned earlier) so there is no need for small files anymore. If needed, data could be written to a single continuous file that doesn't roll over. However, for data management and organization, it is helpful to apply path prefix patterns as described above.

Lastly, a recent addition to the portal when dealing with files in ADLS allows you to use custom delimiters when previewing files, so you can define your own delimiter.

Security will be discussed later in this chapter; the next section will discuss how to work with Azure Data Lake Store using Visual Studio.

Visual Studio

Microsoft provides a number of SDKs and APIs to perform basic operations against Azure Data Lake Store, including .NET, Node, and Java. This section will walk through an example of using the .NET SDK to create a console application to create a folder, upload and download a file, and list files.

Open Visual Studio 2015 and create a new C# console application and give it a name such as ADLSApp or something similar. Once the project is created, open the Package Manager Console by select Tools ➤ NuGet Package Manager ➤ Package Manager Console. In the Package Manager Console, type in and execute the following three commands separately to install the appropriate classes for working with Data Lake Store accounts and filesystem management capabilities:

```
Install-package Microsoft.Azure.Management.DataLake.Store
Install-package Microsoft.Azure.Management.DataLake.StoreUploader -pre
Install-package Microsoft.Rest.ClientRuntime.Azure.Authentication -pre
```

The StoreUploader package provides the capabilities for enabling rapid data uploads into Azure Data Lake Store, and the Authentication package ADAL-based authentication. As of this writing, the StoreUploader and Authentication packages are in preview, thus the -pre parameter. I suggest checking Nuget.org for each of those packages to verify if they are still in preview or not.

Once the three packages have been installed, open Program.cs and add the following statements underneath the existing using statements:

```
using Microsoft.Azure.Management.DataLake.Store;
using Microsoft.Rest.Azure.Authentication;
using Microsoft.Azure.Management.DataLake.StoreUploader;
using Microsoft.Azure.Management.DataLake.Store.Models
```

Next, in the class Program above the Main() method, add the following variable declarations:

```
private static DataLakeStoreAccountManagementClient _adlsClient;
public static DataLakeStoreFileSystemManagementClient _adlsFileSystemClient;

private const string _adlsAccountName = "temperature";
private const string _resourceGroupName = "WestUS-RG";
private const string _location = East US 2";
private const string _subId = "<subscriptionid>";
```

The DataLakeStoreAccountManagementClient class is used to manage aspects of the Azure Data Lake Store. For example, you would use this class to list and manage your Data Lake Store accounts. The DataLakeStoreFileSystemManagementClient class is used to manage aspects of working with the store, such as uploading and downloading files, creating and deleting directories, and so on.

Continuing the example, in the Main method, add the following code. Be sure to add your own Azure subscription ID.

```
string localFolderPath = @"C:\Projects\";
string localFilePath = localFolderPath + "TempSensorData.json";
string remoteFolderPath = "/tempdata/";
string remoteFilePath = remoteFolderPath + "TempSensorData.json";

SynchronizationContext.SetSynchronizationContext(new SynchronizationContext());
var domain = "common";
```

149

```
var nativeClientApp_ClientId = "1950a258-227b-4e31-a9cf-717495945fc2";
var activeDirectoryClientSettings = ActiveDirectoryClientSettings.
UsePromptOnly(nativeClientApp_ClientId, new Uri("urn:ietf:wg:oauth:2.0:oob"));
var creds = UserTokenProvider.LoginWithPromptAsync(domain, activeDirectoryClientSettings).
Result;

_adlsFileSystemClient = new DataLakeStoreFileSystemManagementClient(creds);
_adlsClient = new DataLakeStoreAccountManagementClient(creds);
_adlsClient.SubscriptionId = _subId;

CreateDirectory("tempdata2");
UploadFile(localFilePath, remoteFilePath);
List<FileStatusProperties> myList = ListItems(remoteFilePath);
DownloadFile(remoteFilePath, localFolderPath + "TempSensorData2.json");
```

Reviewing this code for a moment, there are a few things to point out. First, at the time of this writing, Azure Data Lake Store is in preview and is only available in the East US 2 datacenter. Once this service is generally available, it will be available in most if not all Azure data centers.

Next is the section of code that defines the client application's client ID as well as all of the Active Directory settings. You'll notice that a lot of this is hard coded with a client id, domain, and other information. There are two ways to authenticate to Azure Data Lake: end-user authentication and service-to-service authentication. This example uses end-user authentication, which means that an end users identity is verified when trying to interact with ADLS or any service that connects to ADLS. ADLS uses Azure Active Directory (AAD) for authentication, and each Azure subscription can be associated with an instance of AAD. One of the beauties of Azure Active Directory is that it can easily be integrated with an existing on-premises Windows Server Active Directory to leverage existing identity investments.

Anyway, back to this example, the code above uses end-user authentication and thus uses Azure Active Directory with a client ID that is available by default to all Azure subscriptions. It makes walking through this example easier and helps keep the focus on ADLS rather than AAD. However, if you want to use your AAD and application client ID, you need to create an AAD native application and then use the AAD domain, client id, and redirect URI in the ADLS code. For information on creating an Azure Active Directory application, visit https://azure.microsoft.com/en-us/documentation/articles/resource-group-create-service-principal-portal/#create-an-active-directory-application.

The rest of the code above creates a new instance of the DataLakeStoreAccountManagementClient and DataLakeStoreFileSystemManagementClient classes, and then calls methods to create a directory in ADLS, upload a file, list the files in the newly created directory, and then download the file. However, those methods don't exist yet so let's do that. Below the Main method, add the following four methods:

```
public static void CreateDirectory(string path)
{
    _adlsFileSystemClient.FileSystem.Mkdirs(_adlsAccountName, path);
}
public static void UploadFile(string srcFilePath, string destFilePath, bool force = true)
{
    var parameters = new UploadParameters(srcFilePath, destFilePath, _adlsAccountName,
    isOverwrite: force);
    var frontend = new DataLakeStoreFrontEndAdapter(_adlsAccountName, _
    adlsFileSystemClient);
    var uploader = new DataLakeStoreUploader(parameters, frontend);
    uploader.Execute();
}
```

```
public static List<FileStatusProperties> ListItems(string directoryPath)
{
    return _adlsFileSystemClient.FileSystem.ListFileStatus(_adlsAccountName, directoryPath).
    FileStatuses.FileStatus.ToList();
}
public static void DownloadFile(string srcPath, string destPath)
{
    var stream = _adlsFileSystemClient.FileSystem.Open(_adlsAccountName, srcPath);
    var filestream = new FileStream(destPath, FileMode.Create);

    stream.CopyTo(filestream);
    filestream.Close();
    stream.Close();
}
```

If you take a moment and look at this, you'll realize that it is less than 20 lines of code to work with files and directories in ADLS. Most, if not all, of this code needs explanation. Each of these four methods uses the DataLakeStoreFileSystemManagementClient class to easily work with ADLS. The first method simply calls the Mkdirs method to create a directory. The second method uses the class to upload a file to the newly created directory; the third method uses the ListFileStatus method to list all of the files in the newly created directory (should only be one file after initially running the code), and then uses the Open method to download the file.

What you didn't see in any of these examples is the use of the DataLakeStoreAccountManagementClient class even though an instance of it was created. The following code can be added to the project above and simply uses the class to obtain a list of all the Azure Data Lake Store accounts within the specified subscription:

```
public static List<DataLakeStoreAccount> ListAdlStoreAccounts()
{
    var response = _adlsClient.Account.List();
    var accounts = new List<DataLakeStoreAccount>(response);

    return accounts;
}
```

At this point, you should have a fairly solid understanding of Azure Data Lake Store and its capabilities and what you can do with it, but I'll wrap up the chapter discuss the important topic of security and how you can secure data in ADLS.

Security

The benefits of Azure Data Lake Store were discussed earlier in this chapter, and they included unlimited storage capacity for data of any type and size as well as performance and scale. As great as these benefits are, they don't amount to much if the data itself isn't secure. Enterprises spend a lot of time and money to make sure that their vital data is stored securely.

Authentication

Earlier in this chapter I showed one level of security via the implementation of network isolation via firewalls in which IP address ranges are defined so that you can control at the network level who has access to your data. But that is just scratching the surface, because I also discussed how Azure Data Lake Store utilizes Azure Active Directory to manage identity and access of users and groups. Let's look at this topic in more depth.

There are several key benefits of using AAD for access control to ADLS, the biggest one being the ability to simplify identity management. A user or service can be controlled simply by managing the account in the directory. Other benefits include support for multifactor authentication and the ability to authenticate from any client through OAuth and other open standard protocols.

Authorization

Once AAD authenticates a user, ADLS takes over to control permissions within the Data Lake Store. Authorization within ADLS is performed via two different methods to manage both account-related and data-related activities:

- RBAC (Role-based access control) for account management
- POSIX ACL for data access

Role-Based Access Control

There are four basic roles that are defined in ADLS by default, which provide different account operations. The fifth option of assigning no role is also discussed in the following list.

- **No role**: They can use command-line tools only to browse the Data Lake Store.
- **Owner**: A superuser. The Owner has full access to data.
- **Contributor**: Can manage some account management responsibilities, but cannot add or remove roles. Does not allow the management of resources.
- **Reader**: Lets you view everything but you cannot make any changes.
- **User Access Administrator**: Manages user access to accounts.

Each of these roles is available through the Azure portal, PowerShell, and Rest APIs. Also note that not all roles affect data access.

ACLs

ACLs, or access control lists, provide a way to set different permissions for specific named users or named groups. ACLs apply an added security benefit because they provide the ability to implement permissions that differ from the traditional organizational hierarchy of users and groups.

ADLS supports POSIX ACLs. POSIX stands for Portable Operating System Interface for Unix and is a family of standards for maintaining compatibility between operating systems, including Unix. Azure Data Lake Store is a hierarchical file system, much like Hadoop's HDFS and as such, it makes sense for ADLS to support POSIX ACLs.

The ACLs in ADLS control read, write, and execute permissions to resources for the Owner role, the Owners group, and for other users and groups. You can apply ACLs on the root folder, subfolders, and on individual files. To be clear, permissions in the POSIX model implemented and used by ADLS are applied and stored on the item itself and cannot be inherited from a parent item.

Figure 9-5 shows the root temperature folder and subfolders as well as the ACLs applied at the root folder, which so far only apply to me.

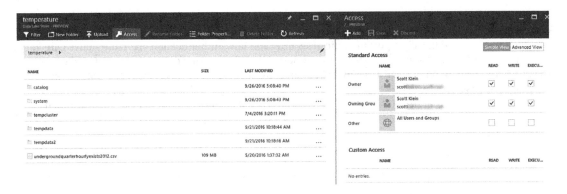

Figure 9-5. Viewing access permissions on the root ALDS folder

A new user can easily be added and their ACLs defined by clicking the New button on the Access blade and selecting the user or group and then selecting their permissions (read, write, execute). See Figure 9-6.

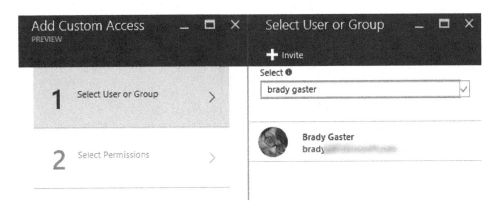

Figure 9-6. Adding a new user and setting permissions

Enterprises expect their business data to be secure, especially in the cloud, and by combining key security capabilities including network isolation, auditing, authentication via AAD, and authorization via ACLs enables Azure Data Lake Store to be an enterprise-ready data repository for big data analytic workloads.

Summary

As enterprises work toward gaining better and deeper insights into their business via big data analytics, they look for a repository aimed at providing the capabilities required for meeting security requirements. This chapter took a look at Azure Data Lake Store, an enterprise-ready cloud service designed for handling big data analytic workloads.

You learned why there's a need for such a service and then you took a look at Azure Data Lake Store, what it is, and where it fits into the spectrum of the Azure Data Lake umbrella. You learned how to create an Azure Data Lake Store via the Azure Portal, and then you looked at many of the aspects and features of Azure Data Lake Store, including the Data Explorer.

From there you looked how to interact with Azure Data Lake Store via the Azure Portal as well as via Visual Studio and the .NET SDK to show how easy it is to work with Azure Data Lake Store.

You then explored the all-important topic of security and how authentication via Azure Active Directory and authorization via access control lists play an important role in securing data in Azure Data Lake Store. You also learned about other security features including network isolation via firewall rules that can be added to further secure the data.

Chapter 10 will look at Azure Data Lake Analytics, a service that enables data analytics via distributed jobs.

CHAPTER 10

■ ■ ■

Azure Data Lake Analytics

Big data analytics is about collecting and analyzing large data sets in order to discover useful information and gain valuable insights previously unknown. The analysis of these large data sets helps uncover hidden patterns, find market trends, discover unknown correlations, and otherwise find treasures of valuable information.

Until recently, the processing of very large data sets was performed on clusters of physically on-premise machines. That meant spending an unnecessary portion of the day managing the distributed infrastructure rather than on focusing on the critical task of data analysis. In reality, big data analytics should be focused on running and managing jobs to understand your precious data, not configuring and tuning hardware clusters.

The shift from on-premises to cloud computing provides a platform for big data analytics in the cloud, commonly called Big Data as-a-service (BDaaS), allowing businesses to focus more on services and getting results rather than spending unnecessary time on infrastructure. This chapter discusses such as service, Azure Data Lake Analytics.

Azure Data Lake Analytics

Azure Data Lake Analytics is a relatively new service, part of the Intelligence and Analytics suite of cloud data services. Azure Data Lake Analytics, or ADLA, exists simply to make big data analytics easy. It does this by letting you focus solely on extracting valuable insights from your data instead of focusing on the hardware and infrastructure management. ADLA also enables the following key capabilities to ensure *analytics is easy*:

- **Scale Dynamically**: ADLA provisions resources dynamically as needed, depending on the amount of data being processed or the compute needed to efficiently complete the job.

- **Unified Query Language**: ADLA includes a new query language called U-SQL, or Unified SQL, which combines the declarative nature of SQL with the expressive power of C#, allowing developers to leverage their existing knowledge and skills.

- **Seamless Integration**: Works seamlessly with Azure data sources and services as well as existing IT systems and infrastructure for security, management, and other investments.

- **Cost-Effective:** Pay on a per-job basis when data is processed. ADLA scales up and down as needed, so you never pay for more than what the job needs.

- **Familiar Tooling**: Visual Studio integration allows you to run, debug, and tune your code and queries in a familiar environment.

© Scott Klein 2017
S. Klein, *IoT Solutions in Microsoft's Azure IoT Suite*, DOI 10.1007/978-1-4842-2143-3_10

Azure Data Lake Analytics is a service designed from scratch with performance, scalability, and easy-of-use in mind. Here is an analytics service that executes over petabytes and exabytes of data and scales dynamically all while requiring no code rewrite or modification.

Management is also made easy through Visual Studio visualizations, which allow you to see how your jobs run at scale and identify performance bottlenecks. This topic will be discussed later in the chapter.

With this brief introduction, this remainder of this chapter will focus on creating and working with Azure Data Lake Analytics.

Creating a New Data Lake Analytics Account

An Azure Data Lake Analytics account can be created via the Azure portal, PowerShell, or any of the supported SDKs such as .NET and Java, as well as the Azure CLI or REST API. There are plenty of options to choose from, but this example will use the Azure portal to create the ADLA account.

Open the Azure Portal by navigating to `Portal.Azure.com` in your favorite browser. Once in the new portal, select New ➤ Intelligence + Analytics ➤ Data Lake Analytics.

The New Data Lake Store blade opens up and, as shown in Figure 10-1, there isn't a whole lot to enter or select to create a Data Lake Analytics account. Simply give it a name; select the appropriate subscription, resource group, location, Data Lake Store, and pricing tier; and then click Create.

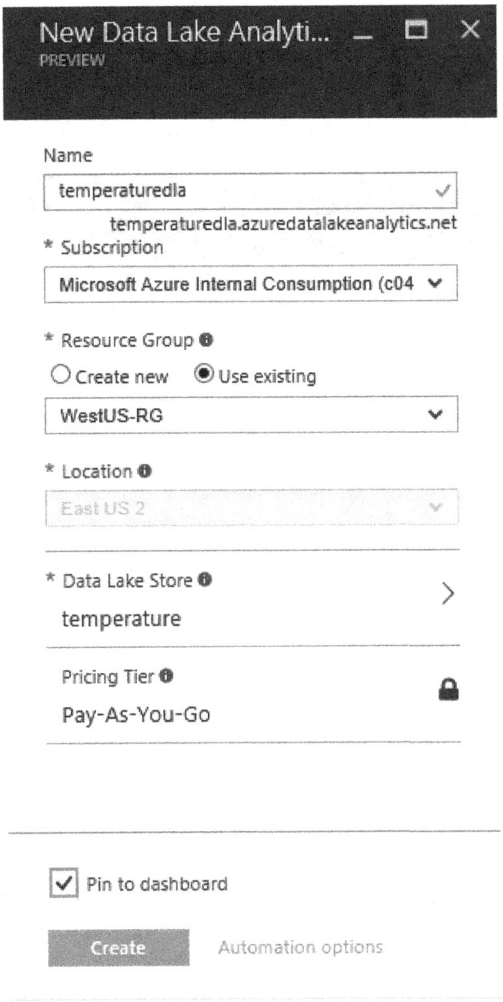

Figure 10-1. *Creating an Azure Data Lake Analytics account*

Now, since this service is in preview as of this writing, the only location available to create an Azure Data Lake Analytics is the East US 2 data center. Many more data centers and regions will be available when this service GAs (is generally available).

In this example, the Data Lake Store that was created in the previous chapter is selected. Otherwise you have the option to create a new Data Lake Store. Also, since the service is in preview, there is no pricing tier to select. Be sure also to select "Pin to dashboard" so you can get to it easily.

The Data Lake Analytics account only takes a minute or two to create, so you won't need to wait long to start working with it. Speaking of which...

Working with Azure Data Lake Analytics

As stated earlier, there are a number of options available in which to create and work with Azure Data Lake Analytics. This section will discuss two of them: the Azure portal and the .NET SDK with Visual Studio. Let's start with the Azure portal.

Azure Portal

Once the Data Lake Analytics account is created, the Overview pane will be presented to you, as shown in Figure 10-2.

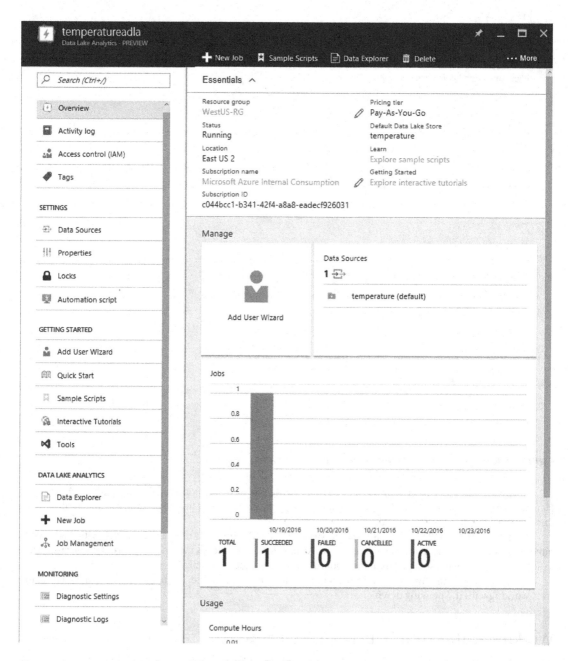

Figure 10-2. Azure Data Lake Analytics Essentials blade

Many of the properties are similar to that of other services, while others are specific to Azure Data Lake Analytics; a few of the latter will be called out here. One of the items to highlight is the Data Explorer link. When the ADLA account was created, one of the options to select was the ADLS account. As shown in Figure 10-2, you can access the contents of the associated Azure Data Lake Store account from within the Azure Data Lake Analytics account.

Also, additional ADLS or Azure Storage accounts can be linked to an ADLA account by clicking the Data Sources item in the Overview pane and adding the appropriate account. You can also set the maximum number of jobs that can run simultaneously, as well as the maximum parallelism for the particular ADLA account by clicking the Properties item in the Overview pane.

Parallelism is defined as the number of compute processes that can happen at the same time. The higher the number, the better the performance, but also an increase in cost. However, parallel nodes in ADLS don't necessarily mean parallel execution. There are several factors that affect parallelism, including file size and how the query is written. Thus, you may have 1,000 nodes but only 300 units of concurrency executing at any one time due to these factors.

Two more items to highlight are the ability to run and manage jobs via the portal. In the Overview pane, click the New Job item to open the New Job pane, shown in Figure 10-3.

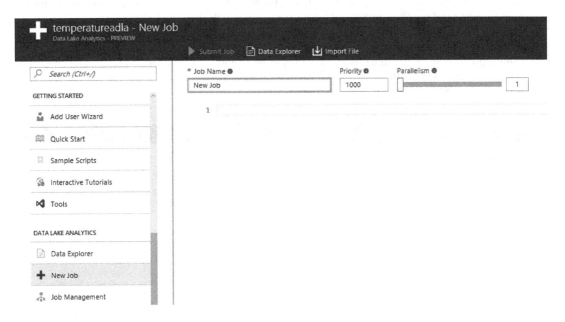

Figure 10-3. *New Job pane*

The New Job pane is where jobs are created and submitted for execution. A job is simply a distributed and parallelized program that transforms and processes data. ADLA receives its instructions from the U-SQL query entered into the ADLA job pane. For example, the following simple U-SQL query reads the data from the source file in Azure Data Lake Store as defined in the FROM clause and then creates a new CSV file as specified in the OUTPUT clause:

```
@searchlog =
    EXTRACT UserId       int,
            Start        DateTime,
            Region       string,
```

```
        Query           string,
        Duration        int?,
        Urls            string,
        ClickedUrls     string
    FROM "/Samples/Data/SearchLog.tsv"
    USING Extractors.Tsv();

OUTPUT @searchlog
    TO "/Output/SearchLog-from-Data-Lake.csv"
USING Outputters.Csv();
```

An ADLA/U-SQL job can be used to execute a lot of different types of data transformation/processing, including ETL, machine learning, analytics, and more. U-SQL, which will be discussed in the next chapter, is simply a way to describe such programs.

In addition to the query to be executed, a job name, priority, and parallelism are also defined for the job. The priority simply sets the importance of the job execution. The lower the number, the higher the priority. If two jobs are awaiting execution, the job with the lower priority will be executed first. Once all the details are specified and set, click the Submit Job button to execute the job.

When a job runs, ADLA determines, based on the priority and parallelism, the number of resources to apply to the job. Behind the scenes, ADLA manages the execution of the job, including the queuing, scheduling, optimization, and actual execution of the job. More about this in Chapter 11.

Once a job completes, the details of the job execution can be viewed by clicking the Job Management node in the Settings blade, as shown in Figure 10-4.

Figure 10-4. *Job Management node*

The Job Management pane by default lists all jobs for the last 30 days. That can be changed by clicking the Filter button and selecting a different time range or creating a custom time range. Additionally, jobs can be filtered by status, job name, author, and more. See Figure 10-5.

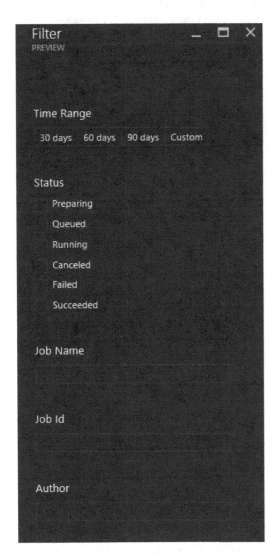

Figure 10-5. *Job filter*

Additionally, as shown in Figure 10-4, clicking a specific job shows the Job Details pane, which also includes a summary of the job as well as the individual steps the job took to complete. In the example above, the priority was 1 with a parallelism of 1, resulting in a 17 second preparation time, 22 second queuing time, and a 29 second job execution time, so overall, just over a minute to run the job. The following section will walk through executing this specific job using Visual Studio to give you an idea of what actually was happening in the job.

The Job Details pane also provides the ability to resubmit the job as well as view the U-SQL that the job executed. If selecting a job that is currently in the Preparing, Queued, or Running status in the Job Management pane, clicking the Cancel Job button will cancel the job and back everything out if needed.

One other item to point out before moving on to Visual Studio: earlier in this chapter I mentioned the ability to access the Data Explorer from within ADLA. The Data Explorer from within ADLA provides a view of the U-SQL catalog not available from within ADLS. The U-SQL catalog is used to structure data and code so that they can be shared by U-SQL scripts to provide the best possible performance with ADLA. As you can see in Figure 10-6, the U-SQL catalog is an Azure SQL Database and was created when the ADLA account was created.

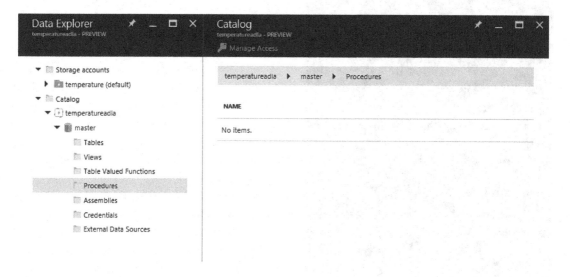

Figure 10-6. *ADLA job catalog*

The Azure portal provides a great interface for working with Azure Data Lake Analytics, from creating, managing, and troubleshooting jobs to exploring data stored in Azure Data Lake Store and other storage mechanisms including Azure Storage.

However, the portal isn't the only place where ADLA can be used, and the following section will show you how to use Visual Studio to create and run a job.

Visual Studio

This section will walk through an example of using Visual Studio to create and submit an ADLA job using the U-SQL snippet earlier in the chapter. The example will simply read a tab-separated value file and use U-SQL to convert it into a comma-separated value file. To follow along, you will need to install the Azure SDK for .NET via the Web Platform Installer.

Before you start cranking out code, there are a couple of things that need to be done. The first item is to go back into the Azure portal, open the Azure Data Lake Analytics account created above, and then open Data Explorer. In Data Explorer, make sure the root node is selected and create a new folder called Samples; within the Samples folder create a subfolder called Data.

Next, go back in the Essentials blade, click the Sample Scripts button and in the Sample Scripts blade, click the Sample Data Missing area, which will copy sample data into your Azure Data Lake Store account. Back in Data Explorer you'll notice 4 files in the Data folder as well as a subfolder called AmbulanceData with about 18 files.

Open Notepad or another text editor and type in the U-SQL from earlier in this chapter and save the file as SampleUSQLScript.txt in C:\Temp. This is the U-SQL that the Visual Studio program will use to execute the analytics job.

Open Visual Studio and create a C# Console application and call it ADLAApp. Once the project is created, open the Package Manager Console because there are a few Nuget packages that need to be installed in order to work with Azure Data Lake Analytics and jobs. If the Package Manager Console is not open, click the Tools menu in Visual Studio and then select Nuget Package Manager ➤ Package Manager Console.

In the Package Manager Console, execute each of the following four lines individually to install the appropriate packages:

```
Install-Package Microsoft.Azure.Management.DataLake.Analytics -Pre
Install-Package Microsoft.Azure.Management.DataLake.Store -Pre
Install-Package Microsoft.Azure.Management.DataLake.StoreUploader -Pre
Install-Package Microsoft.Rest.ClientRuntime.Azure.Authentication -Pre
```

The first Analytics package provides the account, job, and catalog capabilities to work with ADLA. The Store and StoreUploader packages provide the account and filesystem capabilities for enabling rapid data uploads into Azure Data Lake Store. The last package provides ADAL-based (Azure Active Directory Authentication Library) authentication to work with ADLS, ADLA, and other Azure services.

Once the three packages have been installed, open Program.cs and add the following statements underneath the existing using statements:

```
using System.IO;
using Microsoft.Rest;
using Microsoft.Rest.Azure.Authentication;
using Microsoft.Azure.Management.DataLake.Store;
using Microsoft.Azure.Management.DataLake.StoreUploader;
using Microsoft.Azure.Management.DataLake.Analytics;
using Microsoft.Azure.Management.DataLake.Analytics.Models;
```

Next, add the following to the Class Program directly above the Main() method. Be sure to provide your own Azure subscriptionID, ADLA account name, and ADLS account name:

```
private const string SUBSCRIPTIONID = "<subscriptionid>";
private const string CLIENTID = "1950a258-227b-4e31-a9cf-717495945fc2";
private const string DOMAINNAME = "common";

private static string _adlaAccountName = "<ADLAaccountName>";
private static string _adlsAccountName = "<ADLSaccountname>";

private static DataLakeAnalyticsAccountManagementClient _adlaClient;
private static DataLakeStoreFileSystemManagementClient _adlsFileSystemClient;
private static DataLakeAnalyticsJobManagementClient _adlaJobClient;
```

Next, add the following code to the Main() method. If your SampleUSQLScript.txt file is in C:\Temp instead of D:\Temp, be sure to correct that. My file is in D:\Temp, not C:\Temp.

```
string localFolderPath = @"d:\temp\";

// Connect to Azure
var creds = AuthenticateAzure(DOMAINNAME, CLIENTID);

SetupClients(creds, SUBSCRIPTIONID);

// Submit the job
Guid jobId = SubmitJobByPath(localFolderPath + "SampleUSQLScript.txt", "My First ADLA Job");
WaitForNewline("Job submitted.", "Waiting for job completion.");

// Wait for job completion
WaitForJob(jobId);
WaitForNewline("Job completed.", "Downloading job output.");

// Download job output
DownloadFile(@"/Output/SearchLog-from-Data-Lake.csv", localFolderPath +
"SearchLog-from-Data-Lake.csv");

WaitForNewline("Job output downloaded. You can now exit.");
```

Lastly, add the following methods to Program.cs below the Main() method:

```
public static ServiceClientCredentials AuthenticateAzure(string domainName, string
nativeClientAppCLIENTID)
{
    // User login via interactive popup
    SynchronizationContext.SetSynchronizationContext(new SynchronizationContext());
    // Use the client ID of an existing AAD "Native Client" application.
    var activeDirectoryClientSettings = ActiveDirectoryClientSettings.UsePromptOnly
    (nativeClientAppCLIENTID, new Uri("urn:ietf:wg:oauth:2.0:oob"));
    return UserTokenProvider.LoginWithPromptAsync(domainName,
    activeDirectoryClientSettings).Result;
}

public static void SetupClients(ServiceClientCredentials tokenCreds, string subscriptionId)
{
    _adlaClient = new DataLakeAnalyticsAccountManagementClient(tokenCreds);
    _adlaClient.SubscriptionId = subscriptionId;

    _adlaJobClient = new DataLakeAnalyticsJobManagementClient(tokenCreds);

    _adlsFileSystemClient = new DataLakeStoreFileSystemManagementClient(tokenCreds);
}
```

```
public static void DownloadFile(string srcPath, string destPath)
{
    var stream = _adlsFileSystemClient.FileSystem.Open(_adlsAccountName, srcPath);
    var fileStream = new FileStream(destPath, FileMode.Create);

    stream.CopyTo(fileStream);
    fileStream.Close();
    stream.Close();
}

public static void WaitForNewline(string reason, string nextAction = "")
{
    Console.WriteLine(reason + "\r\nPress ENTER to continue...");

    Console.ReadLine();

    if (!String.IsNullOrWhiteSpace(nextAction))
        Console.WriteLine(nextAction);
}

public static List<DataLakeAnalyticsAccount> ListADLAAccounts()
{
    var response = _adlaClient.Account.List();
    var accounts = new List<DataLakeAnalyticsAccount>(response);

    while (response.NextPageLink != null)
    {
        response = _adlaClient.Account.ListNext(response.NextPageLink);
        accounts.AddRange(response);
    }

    Console.WriteLine("You have %i Data Lake Analytics account(s).", accounts.Count);
    for (int i = 0; i < accounts.Count; i++)
    {
        Console.WriteLine(accounts[i].Name);
    }

    return accounts;
}

public static Guid SubmitJobByPath(string scriptPath, string jobName)
{
    var script = File.ReadAllText(scriptPath);

    var jobId = Guid.NewGuid();
    var properties = new USqlJobProperties(script);
    var parameters = new JobInformation(jobName, JobType.USql, properties, priority: 1,
    degreeOfParallelism: 1, jobId: jobId);
    var jobInfo = _adlaJobClient.Job.Create(_adlaAccountName, jobId, parameters);

    return jobId;
}
```

```
public static JobResult WaitForJob(Guid jobId)
{
    var jobInfo = _adlaJobClient.Job.Get(_adlaAccountName, jobId);
    while (jobInfo.State != JobState.Ended)
    {
        jobInfo = _adlaJobClient.Job.Get(_adlaAccountName, jobId);
    }
    return jobInfo.Result.Value;
}
```

A few things to point out before you run the project. First, just like the ADLS example in the previous chapter, there is a section of code that defines the client application client ID as well as all of the Active Directory settings. You'll notice that a lot of this is hard coded with a client ID, domain, and other information. There are two ways to authenticate to Azure Data Lake: end-user authentication and service-to-service authentication. This example uses end-user authentication, which means that an end user's identity is verified when trying to interact with ADLA or any service that connects to ADLA. ADLA uses Azure Active Directory (AAD) for authentication and each Azure subscription can be associated with an instance of AAD.

The code above uses end-user authentication and thus uses Azure Active Directory with a client ID that is available by default to all Azure subscriptions. It makes walking through this example easier and helps keep the focus on ADLS rather than AAD. However, if you want to use your AAD and application client ID, you need to create an AAD native application and then use the AAD domain, client ID, and redirect URI in the ADLS code. For information on creating an Azure Active Directory application, visit https://azure.microsoft.com/en-us/documentation/articles/resource-group-create-service-principal-portal/#create-an-active-directory-application.

Second, the code to create and submit a job is very little. Taking a look at the SubmitJobByPath method, you can see that the path and name of the SampleUSQLScript.txt file is passed to the method as is a job name. The first line of the method reads the contents of the file, and then job information is defined such as the job type, priority, and parallelism, similar to jobs created and executed in the Azure portal.

Notice also that the job type is U-SQL, which means that ADLA should expect a U-SQL script to execute the job. Go back a few pages and look at the U-SQL contained in the SampleUSQLScript.txt file. The U-SQL reads data from the tab-separated value file called SearchLog.tsv that is sitting in Azure Data Lake Store in the Samples\Data directory. It then creates a new comma-separated value file called SearchLog-from-Data-Lake.csv in the Output directory in Azure Data Lake Store. If the Output directory doesn't exist, the job will create it.

U-SQL reads and writes from the two files using extractors and outputters, which provide the ability to define a schema on read and take a rowset and serialize it. More on extractors and outputters in the next chapter.

The job is created and submitted by calling the Create method on the DataLakeAnalyticsJobManagementClient class. Simple as that. The other methods allow the application to check and wait for the job and other supporting methods. Taking a look at just the code needed to create and execute jobs in ADLA, it is very minimal, which shows just how easy it is to work with ADLA jobs.

Go ahead and compile the project to make sure there are no errors. If all compiles ok, run the application. The job will be created and automatically submitted and in about a minute the job will complete. Based on the prompts in the code, you will need to press Enter a couple of times. Once the program exists, you will notice two things: first, the SearchLog-from-Data-Lake.csv in the Output directory in Azure Data Lake Store, and second, the program also downloaded that file to C:\Temp.

From this example, you should understand how to use Visual Studio to create and submit jobs to Azure Data Lake Analytics. You should also have a good understanding of how to use the portal to now go look at the jobs and use the portal to manage jobs. However, we're not quite done because there is some additional information to share which helps write and test U-SQL scripts.

Data Lake Tools

The last section discussed how to use the .NET SDK to work with Azure Data Lake Analytics, that is, to create and submit jobs. However, as good as the SDKs are, they lack the visualization to manage and troubleshoot jobs. This is where the Data Lake Tools for Visual Studio comes into play and solves this problem. The Data Lake Tools for Visual Studio is another mechanism for writing, testing, and submitting jobs to Azure Data Lake Analytics, but also come with the critical ability to visualize aspects of the job.

The Data Lake Tools for Visual Studio can be downloaded from the Microsoft site at

www.microsoft.com/en-us/download/details.aspx?id=49504

I also love the fact that if you don't know the URL, there is a link to them in the Azure portal within the Azure Data Lake Analytics account, as shown in Figure 10-7. This is awesome. The top section has a link to download the Data Lake Tools for Visual Studio, but as you can see there are also links to PowerShell, the Azure CLI, and the .NET SDK.

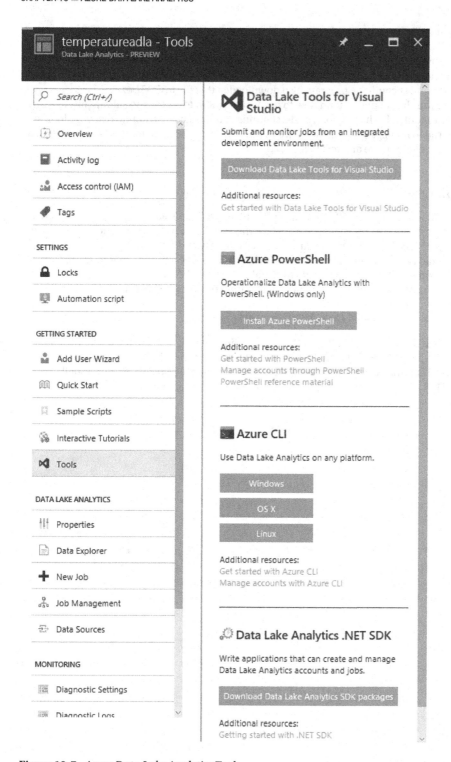

Figure 10-7. *Azure Data Lake Analytics Tools*

Clicking the Data Lake Tools for Visual Studio link takes you to the URL shown above to download the tools. The current release as of this writing is 2.2.2100 dated 10/20/2016 and is the version from which the following example will be taken.

The download contains the install for both Visual Studio 2013 and Visual Studio 2015. The example in this chapter uses 2015 (as do all the other examples in this book). Go ahead and download and run the msi. It doesn't take too long. Make sure Visual Studio is closed during the install.

The Data Lake Tools for Visual Studio does require the Azure SDK version 2.7.1 or higher be installed, which can be installed from the Web Platform Installer. The current version of the Azure SDK for .NET is 2.9.5, which is the recommended version to install. While you're in the Web Platform Installer, if there is an option to install Azure PowerShell, go ahead and do that. You won't be using it in this chapter, but it is good to have nonetheless as you might use it in a later chapter. There are also examples in the Azure documentation on how to use Azure PowerShell to create and submit jobs to Azure Data Lake Analytics.

OK, let's get started. In Visual Studio, create a new project. In the New Project dialog, expand the Azure Data Lake node and select U-SQL (ADLA), and then select the U-SQL Project template, as shown in Figure 10-8.

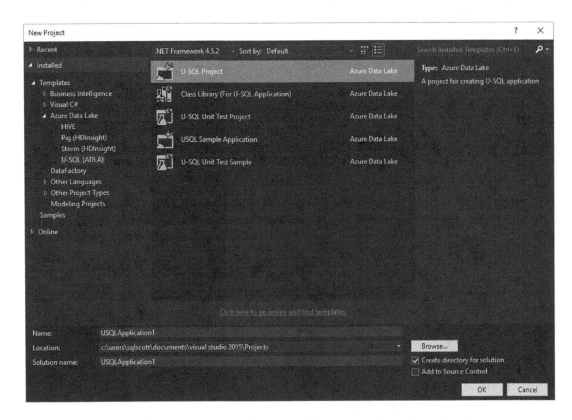

Figure 10-8. *U-SQL project*

Click OK on the New Project dialog. When the project is created, it also creates and displays a `Script.usql` file. In the `Script.usql` file, enter the following:

```
@searchlog =
    EXTRACT UserId        int,
            Start         DateTime,
            Region        string,
            Query         string,
            Duration      int?,
            Urls          string,
            ClickedUrls   string
    FROM "/Samples/Data/SearchLog.tsv"
    USING Extractors.Tsv();

@res =
    SELECT *
    FROM @searchlog;

OUTPUT @res
    TO "/Output/SearchLog-from-Data-Lake.csv"
USING Outputters.Csv();
```

This U-SQL is similar to the U-SQL from earlier in the chapter but differs online in the SELECT statement between the EXTRACT and OUTPUT statements. However, it essentially does the same thing.

There are a few things to point out while in the `Script.usql` and the Data Lake Tools regarding functionality not available in the SDK. While both provide IntelliSense, the Data Lake Tools also supports additional formatting, smart indenting, easy navigation to Azure blob storage paths, and the ability to expand * columns (meaning you can see a list of available columns rather than just SELECT *).

To execute the U-SQL script, in the `Script.usql`, specify the Data Analytics account, database, and schema. Since you are running the U-SQL script from within Visual Studio, you can also compile the script to check for syntax errors. When the script compiles correctly, click Submit in the `Script.usql` window. The job will be executed and the job progress and summary will be displayed as shown in Figure 10-9, similar to what is shown and displayed in the Azure portal.

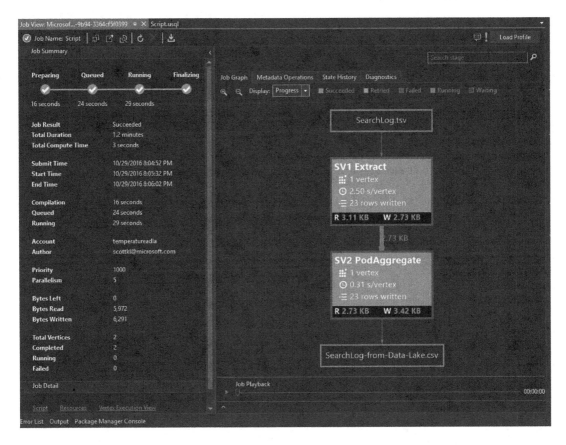

Figure 10-9. *Job results*

The job summary shows the summary information of the current job, including state, progress, submitter, and execution time. The job detail shows the U-SQL script that was executed, the resources used during the execution, and the Vertex execution view (you'll look at this further in the next chapter). The job graph provides visualization into the jobs progress, data read and written, execution time, and other information.

An interesting aspect of the Data Lake Tools is the ability to execute a "job playback," which enables you to visually watch the job execution progress and pinpoint any performance irregularities and bottlenecks. Job playback can be done during and after job execution.

Data Lake Tools also has the ability to run U-SQL scripts locally, which allows you to test U-SQL scripts. To run scripts locally, simply select (Local) in the cluster drop-down list and then click Submit.

Azure Data Lake Tools for Visual Studio is a premium method of working with Azure Data Lake Analytics, authoring and executing U-SQL jobs as well as managing and monitoring jobs to for big data analytics.

Summary

This chapter focused on the three ways to work with Azure Data Lake Analytics, from authoring and executing jobs to managing and monitoring jobs. You first looked at how to use the Azure portal to create an Azure Data Lake Analytics account, followed by working with jobs within the portal.

From there, you learned two ways to author and execute jobs in Visual Studio, first using the .NET SDK followed by the Data Lake Tools. Both provide the functionality to create and execute jobs but the latter provides the additional benefit of visualizing aspects of the job executing to help manager and monitor job activity and execution.

You also got a sneak peek into the U-SQL language that is used by Azure Data Lake Analytics to process all types of data at scale. You'll dive deeper into the U-SQL language in the next chapter.

CHAPTER 11

U-SQL

The last two chapters focused on big data storage and analytical compute services for performing high-scale analytics in Microsoft Azure, Azure Data Lake Store (ADLS), and Azure Data Lake Analytics (ADLA). Azure Data Lake Store is a hyperscale data repository for the enterprise for big data analytic workloads with no limit to file size, ingestion speed, or types of files. Azure Data Lake Analytics lets you focus solely on extracting valuable insights from your data instead of focusing on the hardware and infrastructure management.

Thus, ADLS provides the storage and ADLA provides the analytical compute power. The "glue" that stitches these two services together is the topic of this chapter, the U-SQL language. U-SQL was briefly discussed in Chapter 9 and used somewhat in Chapter 10, but this chapter will focus entirely on the U-SQL language because it is a key component in Azure Data Lake Analytics.

What and Why

The U-SQL language is an analytical processing language built for the sole purpose of processing and analyzing all data at any scale. The "U" in U-SQL stands for *Unified* because it is a language that combines the expressive power of C# with the querying capabilities of T-SQL. The U-SQL language is scalable by design, providing distributed querying capabilities across data sources including Azure Data Lake Store and Azure SQL Database.

To understand the concept and the underlying need for U-SQL, let's back the bus up and revisit, albeit briefly, the challenges of processing and understanding big data. Chapter 1 spent some significant time outlining these challenges, which can be summarized down to the different types and vast amount of data that is being generated today. Structured and unstructured data, and relational and non-relational data, is being generated at a rapid pace today with a great need to process and analyze it.

The aforementioned characteristics of data, even big data, cause great consternation when trying to analyze the data quickly and efficiently. Until U-SQL, processing and analyzing vast amounts of different types of data required a ton of custom coding with the challenge of building in needed performance and scalability. Not easy to do.

Thus, U-SQL was designed from the ground up to overcome the challenges of processing big data with the following goals in mind:

- **Scalability**: With built-in scale-out architecture, U-SQL and Azure Data Lake Analytics will efficiently and easily scale to process any size of data.

- **Extensibility**: By building on the extensibility model of C#, users can add custom code easily to add proprietary logic or algorithms.

- **No limits**: Process any type and size of data.

© Scott Klein 2017
S. Klein, *IoT Solutions in Microsoft's Azure IoT Suite*, DOI 10.1007/978-1-4842-2143-3_11

U-SQL is built on the concept and learnings of SCOPE (Structured Computations Optimized for Parallel Execution), a declarative and extensible scripting language designed to provide massive data analytics easily. SCOPE borrows many of its features from SQL since it deals with data, thus by combining the familiar language semantics of T-SQL with the expressive language of C#, you get a language and an experience in which you can easily write powerful analytical queries.

While the premise of Azure Data Lake Analytics and U-SQL is the ability to process large masses of data at scale, the responsibility is yours to ensure that queries are written properly and smart patterns are used. U-SQL and Azure Data Lake Analytics will process what you feed it, thus understanding the intricacies of use of U-SQL is beyond beneficial.

Michael Rys, the creator of the U-SQL language, said it best: U-SQL unifies both paradigms, unifies structure and unstructured data process, unifies the declarative and custom imperative coding experience, and unifies the experience around extending your language capabilities.

Architecture

There are two key concepts and principles that need to be understood in order to efficiently work with U-SQL and even to fully appreciate the power and flexibility U-SQL has to offer: the core language principles and the job execution lifetime. This section will first take a look at the essential language philosophy and principles needed to write effective U-SQL queries and then follow that up by diving into what happens internally when a U-SQL query is submitted for execution.

Core Language Principles

U-SQL isn't difficult to learn but understanding the basic layout and syntax will be helpful in writing proper queries. U-SQL is a full-featured language, in that, to do meaningful things takes a lot of rationalization of technique, both programmatic and know-how. As you learn U-SQL, you might have the tendency to compare what U-SQL and Azure Data Lake Analytics do to an ETL process. This is a common thought but not a correct one. U-SQL follows a LET process: Load, Extract, Transform. In other words, U-SQL works with data and executes jobs in batches such that the data is preloaded (stored in an Azure Data Lake Store, for example) and then queried by U-SQL, which is when the Extract and Transform is performed.

At its core, the simplest query contains two statements: an EXTRACT and OUTPUT. For example, the following U-SQL query reads data from the SearchLog.tsv file sitting in Azure Data Lake Store and then saves the data back into Azure Data Lake Store into a different file with a CSV format.

```
@searchlog =
    EXTRACT UserId         int,
            Start          DateTime,
            Region         string,
            Query          string,
            Duration       int?,
            Urls           string,
            ClickedUrls    string
    FROM "/Samples/Data/SearchLog.tsv"
    USING Extractors.Tsv();

@res =
    SELECT *
    FROM @searchlog;
```

```
OUTPUT @res
    TO "/Output/SearchLog-from-Data-Lake.csv"
USING Outputters.Csv();
```

There are a few things to point out in the above query. Remember that U-SQL is a combination of T-SQL and C#, so the parser needs to be able to differentiate between the two. As such, there are certain requirements that must be followed. For example, all keywords that look like T-SQL must be in uppercase. Also, the equality operator of "=" in U-SQL is not one, but two. For example,

```
WHERE author == "@SQLScott"
```

Notice the second difference: that string values are wrapped in double quotes in U-SQL, not single quotes. The equality and quoting difference is a C# standard, not a T-SQL standard. Essentially, watch out if you come from a T-SQL background because using the wrong case or valued logic such as dealing with nulls or equality will result in errors.

Extractors, Outputters, and Transformers

Let's talk a minute about extractors and outputters.

- **Extractors**: Obtain data from the source and define schema on read. Datatypes are based on C# datatypes.

- **Outputters**: Write the results to a destination such as a file or a U-SQL table for further processing. This process takes a rowset (equivalent to a row in SQL) and serializes it.

- **Transformers**: Unlike extractors or outputters where there is an EXTRACT and OUTPUT keyword, there is no TRANSFORM keyword. The transformation happens after the EXTRACT and is up to you on how to transform it. More on this later.

U-SQL comes with a number of presupplied and built-in extractors and outputters that work with CSV files, TSF files, and text files. The text file extractor and outputter provide the ability to specify a delimiter parameter, while the CSV and TSV extractor and outputter are derivatives of the text extractor and outputter with the appropriately defined parameter. Additionally, the outputters come with additional parameters to help define items such as encoding, how to deal with nulls and escapes, and dateTime formats.

With this information in mind, let's take a deeper look at the query as a whole. It's a fairly simple query. The EXTRACT statement simply allows you to schematize the unstructured data without needing to create a metadata object for it. This allows U-SQL to declare the format of the data being read from the file in a "schema on read" application. You can, in fact, have more than one EXTRACT statement and more than one OUTPUT statement in a U-SQL query.

U-SQL works with batches, or rowsets. Thus, the EXTRACT statement is assigned a variable of @ searchlog. At this point, the entire contents of searchlog.tsv are not assigned to @searchlog; it is simply a reference, much like a common table expression in T-SQL.

The columns in the EXTRACT statement must include all of the columns in the file; otherwise an error is returned. The filtering of the data is done in another statement. In the example above, it is done in the very next statement, the SELECT. In this example, you are selecting all the rows and columns, but you could easily change the query to the following to provide filtering and allow the selection of specific columns (following standard ANSI patterns):

```
@res =
    SELECT Region, Query, Urls, ClickedUrls
    FROM @searchlog
    WHERE UserId == "@SQLScott";
```

At this point, the @res rowset contains the results of your transformation. Assigning each expression to a variable allows U-SQL to incrementally transform and combine data step by step.

Next, the OUTPUT keyword informs U-SQL of what you want to do with the results of the transformation, which in this case is to store it back in Azure Data Lake Store. In the example above, the results contained in @res are output to Azure Data Lake Store in a different directory and filename. Notice that it uses the CSV outputter to save the results as comma-separated. This is done by applying the USING statement and specifying which outputter to use.

Going back to the transformation for a second, the transformation happens between the EXTRACT and OUTPUT and is essentially done via a SQL statement or via a custom user-defined operator.

Let's use the U-SQL query below to illustrate a lot of what we have been discussing. Notice that the first two statements create a table and define a reference to an assembly. I'll come back to that later on. Following that, though, are two EXTRACT statements, followed by a nice example of a transformation statement where you join the two rowsets from the EXTRACT statements and then write the results of the transformation to both a text file and the table created earlier.

```
REFERENCE MyDB.MyAssembly
CREATE TABLE T (cid int, first_order DateTime, last_order DateTime, order_count int, order_
amount float);

@o = EXTRACT oid int, cid int, odate DateTime, amount float
    FROM "/input/orders.txt"
    USING Extractors.Csv();

@c = EXTRACT cid int, name string, city string
    FROM "/input/customers.txt"
    USING Extractors.Csv();

@j = SELECT c.cid,
        MIN(o.date) AS firstorder,
        MAX(o.date) as lastorder,
        COUNT(o.oid) AS ordercnt,
        AGG<MyAgg.MySum>(c.amount) AS totalamount
    FROM @c AS c LEFT OUTER JOIN @o AS o ON c.cid == o.cid
    WHERE c.city.StartsWith("New") && MyNamespace.MyFuction(o.odate) > 10
    GROUP BY c.cid;

OUTPUT @j TO "/output/result.txt"
USING new MyDate.Write();

INSERT INTO T SELECT * FROM @j;
```

To help add clarity to all of this, let's look at this from language philosophy perspective. Given the declarative and transformational capabilities in U-SQL, let's take a look at the U-SQL query above through the lenses of the U-SQL language philosophy. Using the chart below, apply the numbered coding to the subsequent query.

U-SQL Capabilities

Operates on Unstructured and Structured Data	
Schema on read over files	1
Relational metadata objects	2
Extensibility Features	
Type system based on C#	3
C# Expression Language	4
User-defined functions (U-SQL and C#)	5
User-defined Aggregators (C#)	6
User-defined Operators (UDO) (C#)	7

```
REFERENCE MyDB.MyAssembly <2>
CREATE TABLE T <2> (cid <2> int <3>, first_order DateTime, last_order DateTime, order_count
int, order_amount float);

@o = EXTRACT oid int, cid int, odate DateTime, amount float <3>
    FROM <3> "/input/orders.txt" <1>
    USING Extractors.Csv();<3>

@c = EXTRACT cid int, name string, city string <3>
    FROM <3> "/input/customers.txt" <1>
    USING Extractors.Csv(); <3>

@j = SELECT <3> c.cid,
        MIN <3> (o.date) AS firstorder,
        MAX <3> (o.date) as lastorder,
        COUNT <3> (o.oid) AS ordercnt,
        AGG<MyAgg.MySum> <6> (c.amount) AS totalamount
    FROM @c <3> AS c LEFT OUTER JOIN <3> @o AS o ON c.cid == <4> o.cid
    WHERE c.city.StartsWith("New") <4> && <4> MyNamespace.MyFuction <5> (o.odate) > 10 <4>
    GROUP BY <3> c.cid;

OUTPUT @j TO <3> "/output/result.txt" <1>
USING new <3> MyDate.Write(); <7>

INSERT INTO T SELECT * FROM <2> @j <3>;
```

Hopefully this helps provide insight into how U-SQL is structured to follow a LET pattern of retrieving data, transforming the rowsets, storing the transformed rowsets, and a breakdown of a U-SQL statement. To further this discussion, the following section will look at what happens when a U-SQL query is submitted for execution.

U-SQL Batch Job Execution Lifetime

So, what really happens when a U-SQL query is submitted? Figure 11-1 shows the steps and phases that take place within Azure Data Lake Analytics during the execution of a U-SQL query.

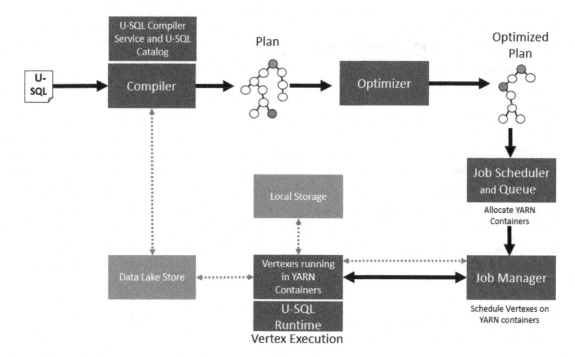

Figure 11-1. *U-SQL job execution lifetime*

Beginning on the top left, the U-SQL query is submitted and is picked up by the compiler to generate an appropriate execution plan. As part of plan generation, the compiler will communicate with Azure Data Lake Store to determine and figure out pertinent and important information, such as, do the files exists, how much data is there, how big are the files, etc.

Once the compiler has accumulated all of this information, it generates a plan and outputs that plan, which is then picked up by the optimizer. The optimizer looks at the plan and corresponding information, and looks at a host of optimizations in order to produce the best optimized execution plan. The output from the optimizer is this optimized plan.

The optimized plan is then picked up by the job scheduler and queue, whose job it is to look at the optimized plan and figure out a plan for the YARN containers in order to execute the job. These YARN containers are not docker containers; they are virtual machines dedicated to the execution of ADLA jobs. ADLA will allocate YARN containers on virtual machines. The job scheduler figures how many YARN containers are needed and will schedule the job appropriately based on the availability of VMs and YARN containers, and coordinate that work with the job manager. There is not necessarily a one-to-one mapping between YARN containers and virtual machines. YARN containers generally have a small memory overhead of up to 300MB so the more containers or tasks you have, the more the subsystem consumes resources.

The job manager, which in fact itself takes a YARN container (which you do not pay for), picks up the plan from the job scheduler and deploys the pieces of work from the job scheduler to the YARN containers and then orchestrates their execution, which includes deploying the U-SQL code, the U-SQL runtime, etc.

A vertex is a unit of work within Azure Data Lake Analytics. A vertex receives its work from the job manager. It reads its data from the Data Lake Store but all intermediate work happens on local disk for performance; all data will eventually will written back to store. When the last vertex is done, the job is done and the job manager declares success.

U-SQL scripts are parsed by the single engine, the U-SQL compiler, the output of which is C++ and C#. The C++ and C# compilers then compile the C++ and C# into unmanaged code, which is then run with the U-SQL runtime to execute the script on the vertices.

We can summarize the image this way; everything on the top of the image is the preparation and queueing phase. All the steps along the bottom are in the execution and finalization phases.

Extensibility

A key advantage and one of the main design goals of U-SQL is the ability to extend its capabilities by adding customer user code in the form of .NET assemblies. These assemblies can then be used and referenced in a U-SQL script via the REFERENCE ASSEMBLY statement. This section will walk through the various extensibility methods for U-SQL. U-SQL can be extended through the following means:

- User-Defined Functions

- User-Defined Aggregators

- User-Defined Operators

- Inline C# Expressions

Before you get started, let's quickly discuss the concept of the U-SQL catalog and associated database. As mentioned, extending U-SQL requires the creation of assemblies and objects to be referenced in U-SQL scripts. For these objects to be shared, a mechanism is required to make sharing and securing these objects easy and efficient in terms of dealing with the objects and accessing and working with data. Thus, the idea and use of a catalog.

Every Azure Data Lake Analytics account contains exactly one U-SQL catalog. The catalog cannot be deleted or shared between ADLA accounts. Each catalog by default contains a single database named Master. The Master database also cannot be deleted but additional databases can be created. By default, all assemblies are created in the Master database unless you specify a different database. Just like T-SQL, you switch between databases using the USE statement.

Best practice states that additional databases be created. Thus, before you get started extended U-SQL, the following section will create an additional database in the U-SQL catalog.

In Visual Studio, open Server Explorer, expand the Azure node, and then expand the Data Lake Analytics node. If your ADLA account is not listed, right-click the Azure node and select Refresh from the context menu.

If it still does not show up, you will need to connect to your Azure subscription. Right-click the Azure node and select "Connect to Microsoft Azure Subscription." Authenticate to your subscription by supplying the username and password when prompted. Once authenticated, create a new U-SQL project, as shown in Figure 11-2.

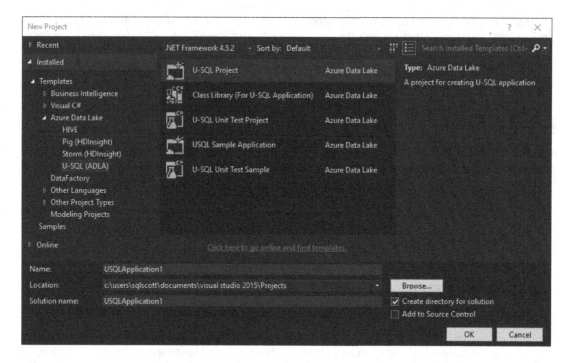

Figure 11-2. Creating a U-SQL project

The U-SQL project templates are installed as part of the Microsoft Azure SDK, which is installed via the Web Platform Installer. The Web Platform Installer can be downloaded and installed from `www.microsoft.com/web/downloads/`.

Download the Web Platform Installer and install the latest Azure SDK for .NET (you may need to exit Visual Studio). Once installed, restart Visual Studio and create a new U-SQL project.

As part of the project creation, a new U-SQL script called `Script.usql` will be created automatically and opened. In the script window, type `CREATE DATABASE IF NOT EXISTS Temperature`, as shown in Figure 11-3.

Before clicking submit, make sure the appropriate Azure Data Lake Analytics account is selected. The drop-down box to the immediate right of the Submit dropdown/button may default to (local). Click the drop-down box and select your ADLA account. The database drop-down, the second drop-down from the right, will default to the Master database. That is OK; leave it as is. As well, leave the dbo drop-down as is because you want to execute this script in that schema.

Figure 11-3. Creating an Azure Data Lake Analytics Database

If you are thinking that the database and schema are a lot like SQL Server, that is because they are. The database, whether it be `Master` or another database you create, contains assemblies, table-valued functions (UDFs), tables, and schemas.

Once you have everything configured, click the Submit button. Immediately, the Job Summary tab will display, showing you the process of the job. Yes, even creating a database is submitted to ADLA as a job. Thus, this U-SQL script follows the same job flow explained earlier, like any other U-SQL script submitted.

You will see in the job summary, shown in Figure 11-4, that the job is being prepared, queued, executed, and finalized, following the same steps as outlined above. It also shows how long the job took to complete and other information. Creating a database isn't too elaborate so the information for this job will be minimal.

Going back to Server Explorer, right-click the U-SQL Databases node and click Refresh in the context menu. The new database will now be listed, as shown in Figure 11-4.

Now, you could just have easily right-clicked the U-SQL Databases node in Server Explorer and clicked Create Database from the context menu. But what is the fun in doing that? Honestly, you'll get the same Job Summary screen, but this example used a script in order to explain the toolbar and options available. You will need to know this when you start authoring U-SQL scripts.

Go ahead and expand the Temperature database node in Server Explorer and take a look at the subfolders. You see things like assemblies, credentials, procedures, schemas, tables, and more. These are all different types of objects that can be created and used in your U-SQL script.

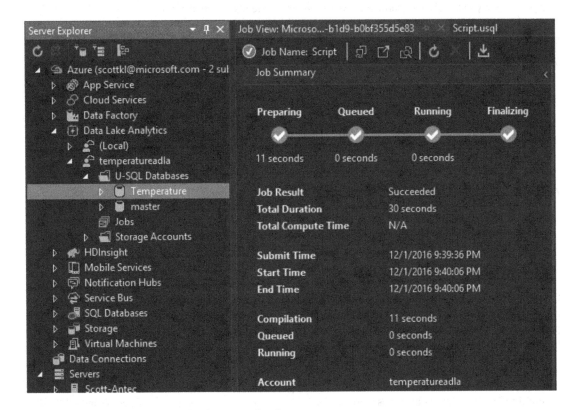

Figure 11-4. Job execution creating the ADLA database

Briefly, switch back to the Azure portal and open your Azure Data Lake Analytics account. Click the Data Explorer link and take a moment to examine the Data Explorer blade. In this blade, you see the U-SQL catalog (which is the same name as the ADLA account) as well as the databases within the account, similar to Figure 11-5.

Figure 11-5. *The ADLA Temperature database listed in the Azure portal*

As a quick note, since the database is a regular database, access to it can be defined via permissions defined in the portal. As shown in Figure 11-6, selecting the database and then clicking the Manage Access button will give you the ability to add/remove users and define read or read/write access.

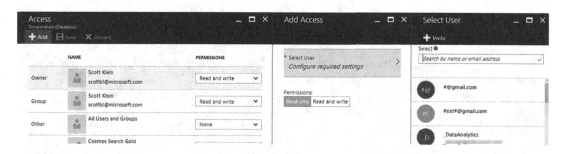

Figure 11-6. *Configuring permission on the ADLA Temperature database*

Let's now look at extending U-SQL by addressing the topic of adding user-specific code.

Assemblies

U-SQL provides the ability to use .NET assemblies to extend its functionality and add additional complex processing. Creating and registering assemblies allows others to use the code in their scripts and is the preferred way to manage user-defined functions and operators. This section will discuss the different ways that U-SQL can be extended to further its processing capabilities.

The best and easiest way to create and use assemblies is via the Data Lake Tools for Visual Studio, which you installed earlier. Open up Visual Studio and from the list of installed templates select the U-SQL (ADLA) option and then select the Class Library (for U-SQL Application), as shown in Figure 11-7. As it states in the description, the U-SQL class library is a project for creating a custom code which can be compiled into a .dll which can be accessed in U-SQL.

Earlier in the chapter you used the same U-SQL (ADLA) option but selected the U-SQL Project to write and execute U-SQL queries and jobs. In this case, though, you'll be creating a simple class library which will then be registered as an assembly so that anyone writing U-SQL can access and use the library.

In the New project dialog, select the Class Library (for U-SQL Application) option and click OK. For this example, I kept the default project name.

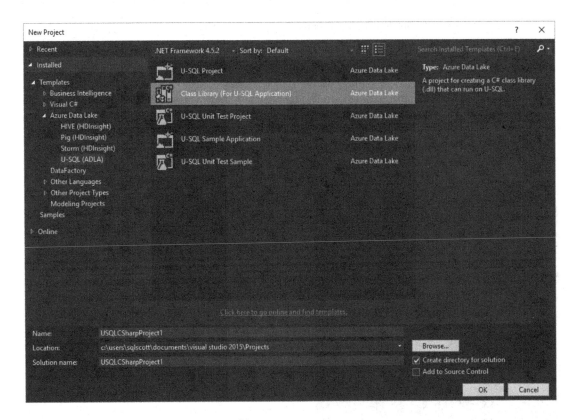

Figure 11-7. *Creating a U-SQL .NET assembly*

When the project is created, open the Solution Explorer window and expand the References node for the project. Notice that it contains references to two name spaces: Microsoft.Analytics.Interfaces and Microsoft.Analytics.Types. The Interfaces assembly is required when creating and defining UDT (user-defined type) interfaces and the Types assembly allows the usage of standard C# and SQL expressions.

The project is created with a default class, Class1.cs, and opening the class is similar to a standard C# class in that all you need to do is define the methods within the class. Thus, a simple method to split incoming temperature data could look like the following:

```
using Microsoft.Analytics.Interfaces;
using Microsoft.Analytics.Types.Sql;
using System;
using System.Collections.Generic;
using System.IO;
using System.Linq;
using System.Text;

namespace USQLCSharpProject1
{
    public class Class1
    {
        public static SqlArray<string> get_temps(string tempdata)
        {
```

183

```
            return new SqlArray<string>(tempdata.Split(new char[] { ' ', ',' }));
        }
    }
}
```

Once the method is defined and the solution is compiled into a .dll, registering the assembly is fairly simple. Connect to your Azure Data Lake account via the Server Explorer window in Visual Studio. Expand the Data Lake Analytics code, expand the node for your specific Data Lake account, and expand the U-SQL Databases node. Right-click the Master database and select Register Assembly from the context menu, as shown in Figure 11-8.

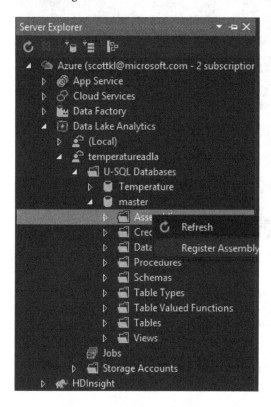

Figure 11-8. *Registering the assembly*

When registering assemblies, they are registered in U-SQL's metadata catalog, which you learned about earlier. Registering assemblies in the Master database makes them available everyone, but you can add a layer of security by registering them in a specific database, which is a good way to organize objects for specific groups or organizations. Once the assembly has been registered, it can be used in U-SQL scripts as follows:

```
@m = SELECT USQLCSharpProject1.Class1.get_temps(temps) AS temps FROM @t
```

User-Defined Functions, Aggregators, and Operators

Another one of the ways to extend U-SQL is through user-defined objects such as functions, aggregators, and operators.

Functions

User-defined functions are a great way to include complex functionality into your custom code beyond the standard or typical C# expression language, such as recursion or procedural logic. The example above can be considered a user-defined function, but you could also do something more procedural, such as the following:

```
namespace TemperatureProject
{
    public class UDFs
    {
        public static double ConvertFahrenheitToCelsius(double temp)
        {
            return (9.0 / 5.0) * (temp - 32);
        }

        public static double ConvertFahrenheitToKelvin(double temp)
        {
            return (((temp - 32) / 1.8) + 273.15);
        }
    }
}
```

The code you wrote way back in Chapter 2 sends the temperature data as Fahrenheit. The code above contains two user-defined functions: one converts the Fahrenheit temperature back to Celsius and the other converts the Fahrenheit temperature to Kelvin. Realistically, you would have 9 UDFs in the assembly (a 3x3 matrix) because depending on the type of temperature you could convert it to one of the others.

As such, you can call the UDF as follows:

```
@searchdata =
    EXTRACT Device  string,
            Sensor  string,
            Temp    double,
            Humidity    double,
            Time    DateTime
    FROM "/tempcluster/logs/2017/01/tempdata.csv"

@t =
    SELECT Device,
        TemperatureProject.UDFs.ConvertFahrenheitToCelsius(Temp) AS value
    FROM @searchdata
```

In the code, you first gather all of the temperature data from the data in Azure Data Lake Store into the @searchdata variable. You then query the device and, using the UDF, the converted temperature using the passed-in temperature.

Another interesting UDF would be to create a similar UDF that takes the temperature as a parameter, and returns a true/false value depending on certain criteria. Meaning, is the temperature within the allotted temperature bounds?

Operators

User-defined operators are U-SQL's custom-coded rowset operators. Since they are written in C#, they can generate, process, and consume rowsets. An example of a UDO (user-defined operator) is the following:

```
[SqlUserDefinedProcessor]
public class TempProcessor : IProcessor
{
    //return the device and valid temperature
    public override IRow Process(IRow input, IUpdatableRow output)
    {
        string Device = input.Get<string>("Device");
        int Temp = input.Get<int>("Temp");
        string value = Device + "-" + UDFs.isOutofBounds(Temp);
        output.Set<string>("Device", value);
        return output.AsReadOnly();
    }
}
```

In this example, the code takes the device name and temperature and then, using the earlier UDF, combines the device name with the true/false value returned from the UDF based on the temperature passed in. With that, you can call the UDO as follows:

```
@searchdata =
    EXTRACT Device   string,
            Sensor   string,
            Temp     int,
            Humidity    int,
            Time     DateTime
    FROM "/tempcluster/logs/2017/01/tempdata.csv"

@t =
    PROCESS @searchdata
    PRODUCE Device string,
            Sensor string,
            Time DateTime
    READONLY Sensor, Time
    REQUIRED Device, Temp
    USING TemperatureProject.UDFs.TempProcessor();
```

As in the first example, you first gather all of the temperature data from the data in Azure Data Lake Store into the @searchdata variable. You then query the device, sensor, and, using the UDO, a value, which returns the device name combined with a true/false value which, as explained above, came from the UDF.

Aggregators

UDAs (user-defined aggregators) enable custom aggregation logic, such as GROUP BY clauses, to be plugged in to U-SQL's aggregation processing. An example of a UDA is the following:

```
public class TempAggregate : IAggregate<string, int, int>
{
    int count;
```

```
public override void Init()
{
    count = 0;
}

public override void Accumulate(string device, int temp)
{
    if (device == "Tessel" && temp >= 80)
    {
        count += 1;
    }
}

public override int Terminate()
{
    return count;
}
}
```

In this example of a user-defined aggregator, the class is initialized with the count variable set to 0. Each time the Accumulate is called, the device and temperature are passed in. If the device name is Tessel and the temperature is above or greater than 80 degrees Fahrenheit, add 1 to the count. In this example, you want to know in real time how many of the Tessel devices are reporting high temperatures (again, a fairly simple example of how UDAs can work). With that, you can use the UDA as follows:

```
@searchdata =
    EXTRACT Device   string,
            Sensor   string,
            Temp     int,
            Humidity     int,
            Time     DateTime
    FROM "/tempcluster/logs/2017/01/tempdata.csv"

@count =
    SELECT Device,
    AGG<TemperatureProject.UDFs.TempAggregate>(Device, Temp) AS count
    FROM @searchdata
```

Again, you first gather all of the temperature data from the data in Azure Data Lake Store into the @ searchdata variable. You then query the device and count of bad temperatures where the device is a Tessel using the UDA.

Another interesting aggregator would be to pass in the temperature and time and, using that data, do a plot graph to track the temperature over a period of time, and then compare that to data from similar sensors. I'll also mention here that you can also use Map/Reduce in SQL with the IReducer interface. It might be fun to experiment using this interface within a U-SQL extensibility mechanism just to play around with.

Just to be clear, I'm not picking on the Tessel. This is just an example and I highly encourage you to experiment with user-defined functions, operators, and aggregates.

Inline Expressions and Code-Behind

Code-behind is a great way to quickly add custom code into U-SQL, but there are a couple of disadvantages to using code-behind. First, the custom code gets uploaded every time the script is submitted and executed. Second, any code-behind cannot be shared with other U-SQL scripts and jobs.

As such, the preferred method is to copy your code into a U-SQL class library, and then compile and register it as a U-SQL assembly. If you want to limit who has access to the assembly, you can register it within a specific ADLA database.

To illustrate how code-behind works, open up the U-SQL Application project created earlier in this chapter. With the `Script.usql` file open, open up the Solution Explorer and pin the window so it stays. While in Solution Explorer, expand the `Script.usql` node so that it shows `Script.usql.cs`. Open the `Script.usql.cs` file and then dock it next to the `Script.usql` window, so that all three windows (`Script.usql`, `Script.usql.cs`, and Solution Explorer) are side by side, as shown in Figure 11-9.

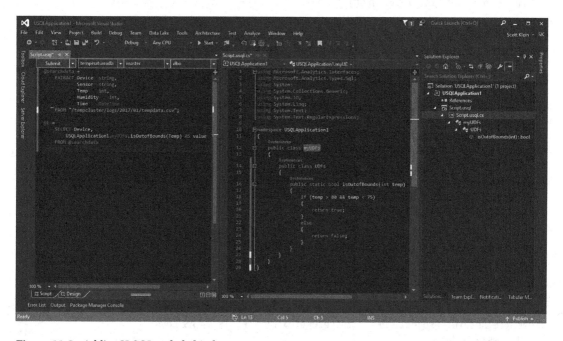

Figure 11-9. *Adding U-SQL code-behind*

The `Script.usql.cs` should have the shell of a class with no methods. Thus, let's take the UDF from the earlier example and paste it here. The `Script.usql.cs` should now look like the following:

```
using Microsoft.Analytics.Interfaces;
using Microsoft.Analytics.Types.Sql;
using System;
using System.Collections.Generic;
using System.IO;
using System.Linq;
using System.Text;
using System.Text.RegularExpressions;
```

```
namespace USQLApplication1
{
    public class myUDFs
    {
        public class UDFs
        {
            public static bool isOutofBounds(int temp)
            {
                if (temp > 80 && temp < 75)
                {
                    return true;
                }
                else
                {
                    return false;
                }
            }
        }
    }
}
```

Next, modify the Script.usql to query the data and call the new function, just like you did earlier:

```
@searchdata =
    EXTRACT Device  string,
            Sensor  string,
            Temp    int,
            Humidity    int,
            Time    DateTime
    FROM "/tempcluster/logs/2017/01/tempdata.csv";

@t =
    SELECT Device,
        USQLApplication1.myUDFs.isOutofBounds(Temp) AS value
    FROM @searchdata
```

Save all the files. You will now see in Solution Explorer, as you continue to expand the Script.usql.cs node, the UDF and functions within the assembly. Now, I'll remind you of the two disadvantages to using code-behind. First, the custom code gets uploaded every time the script is submitted and executed. Second, any code-behind cannot be shared with other U-SQL scripts and jobs. So, keep this in mind.

Now, a quick word on inline expressions. These are expressions imbedded directly in the U-SQL code, such as

```
@t =
    SELECT new SQL.ARRAY<string>(temp.Split(' ').Where(x => x.StartsWith("Tessel"))) AS values
    FROM @searchdata
```

Kind of a cool example of using C# in the middle of SQL. The built-in type `SQL.Array<T>` is a C# object type that makes the SQL/Hive capabilities available. Very cool. This chapter has just scratched the surface with U-SQL, but hopefully provided enough information for you to start digging in and playing with it. As I stated in the introduction of this book, you can find more U-SQL examples on my blog. To get started with U-SQL, this is a great place to start: `https://docs.microsoft.com/en-us/azure/data-lake-analytics/data-lake-analytics-u-sql-get-started`.

Michael Rys's blog is also a fantastic place to learn about U-SQL:

`https://blogs.msdn.microsoft.com/mrys/`

Lastly, the following hands-on lab is a great introduction to working with U-SQL:

`http://aka.ms/usql-hol`

So, have at it! Dig in, and enjoy.

Summary

U-SQL can, at first, seem a bit intimidating, but it doesn't need to be. If you know C# and T-SQL, you are already half way there. The intent of this chapter was to expose you to this wonderful language and its capabilities. You began by learning why U-SQL exists, and then you looked at the architecture and structure as a foundation for understanding its language philosophy. Then you looked at example U-SQL queries and their structure, and from there you looked at U-SQL's batch job execution lifetime and what happens when a U-SQL query is submitted. Understanding this process will help you troubleshoot job execution and performance issues.

From there, you dove into extending U-SQL to add custom code through the use of .NET assemblies and user-defined functions, operators, and aggregates. Finally, you learned how to implement inline expressions and use Visual Studio's code-behind capabilities.

The next chapter discusses Azure HDInsight, Microsoft's Hadoop as a service.

CHAPTER 12

■ ■ ■

Azure HDInsight

Chapter 9 and 10 covered two cloud services in Microsoft Azure that provide hyperscale, distributed storage and analytics for big data workloads: Azure Data Lake Store and Azure Data Lake Analytics. As a refresher, Azure Data Lake Store (ADLS) is the storage mechanism for storing data of any size and type, with the ability to scale appropriately as increased demand and ingestion speed increases. Azure Data Lake Store (ADLA) is an analytics service that lets you focus on gaining valuable data insights via writing and running jobs rather than spending time on the infrastructure.

Chapter 11 talked about the new U-SQL language, part of the Azure Data Lake Analytics service for big data processing. Providing scalability and extensibility without limits on type and amount of data, U-SQL is designed to process and analyze large amounts of data. Together, U-SQL and Azure Data Lake Analytics provide unparalleled big data processing, letting you focus on extracting valuable insights without the need to focus on the hardware and infrastructure management components.

This chapter will focus on Azure HDInsight, the fully-managed Hadoop cloud service and its ecosystem of components and utilities. There are definitely some high-level similarities when you compare ADLA and HDInsight. Both ADLA and HDInsight use a distributed architecture for processing big data, for example. There are other similarities that I will discuss shortly.

The question then becomes one that you should be asking yourself right now if not over the next few pages: why would you choose to use Azure Data Lake Analytics over HDInsight and Hadoop, or vice versa? This is a great question and one that will be answered shortly. But first, let's back up and first cover the "what."

What Is HDInsight?

Simply, HDInsight is Hadoop as a Service. Instead of installing and configuring an on-premises big data Hadoop infrastructure, Azure HDInsight provides a pay-as-you-go solution for Hadoop-based big data batch processing. Azure HDInsight utilizes a cluster of Azure virtual machines running the Hortonworks Data Platform (HDP) and uses the Hadoop Distributed File System (HDFS), Azure Blob Storage (wasb/wasbs), or Azure Data Lake Store as integrated data sources.

An Azure HDInsight cluster can be created in a matter of minutes, and the cluster can be deleted just as quickly, thus the pay-as-you-go model. Because of the data source integration with Azure Blob Storage and Azure Data Lake Store, when the cluster is deleted, your data is still persisted.

Azure HDInsight is not only a fully-managed cloud Hadoop offering, but HDInsight also provides optimized open source clusters for other Hadoop cluster components such as Spark, Hive, Storm, Kafka, and more. However, to understand the difference between Azure Data Lake Analytics and HDInsight, let's spend a few minutes talking about Hadoop.

© Scott Klein 2017
S. Klein, *IoT Solutions in Microsoft's Azure IoT Suite*, DOI 10.1007/978-1-4842-2143-3_12

What Is Hadoop?

Hadoop is a set of open-source components, written in Java, designed for the distributed processing and analysis of large data sets across one or many computer clusters. Azure HDInsight is based on the open source Hortonworks Apache Hadoop Platform distribution base, providing a scalable and fault-tolerant big data processing and analytics solution. Hadoop provides many of the tools necessary to query, transform, and analyze data, including

- **Hive**: Using a SQL-like language called HiveQL, Hive provides the ability to overlay a schema onto a table when executing a query. Schema is not static and is applied on read.

- **Storm**: A real-time, distributed system for quickly and efficiently processing large streams of data. Good for real-time analytics.

- **Pig**: Using a procedural language called Pig Latin, Pig lets you create schemas and execute queries through scripts.

- **MapReduce**: Written in Java and executed by the Hadoop framework, MapReduce is a framework in which applications can be written to process large amounts of data in parallel. Map takes a set of data and breaks it down into tuples (key/value pairs). The Reduce part takes the output of the Map job, combining all the tuples into a smaller set of tuples. This process makes scaling over multiple nodes much easier.

The idea with Hadoop is that it is designed to scale from a single node to multiple nodes, thousands if needed, with each node in the cluster offering its own processing, computation, and storage. Thus, the ability to execute map/reduce jobs effectively and efficiently. Hadoop is designed for high availability such that Hadoop itself can detect and handle failures and reroute data and jobs when necessary.

Hadoop includes the Hadoop Distributed File System, which is a highly-distributed file system with built-in fault tolerance and high throughput. It is designed to process very large datasets, typically gigabytes to terabytes in size. Hadoop clusters can range from tens of nodes to hundreds of nodes or more, thus HDFS should be able to process and handle millions of files.

Hadoop also includes a framework called YARN which provides job scheduling and cluster resource management. This framework is utilized by the MapReduce module for the parallel processing of data.

Thus, when you spin up an Azure HDInsight Hadoop cluster, you are getting all the benefits of Hadoop described above. The only difference, as discussed earlier, is that HDInsight is fully managed without the worry of installing and configuring an on-premises big data Hadoop infrastructure. And, since you only pay for the resources you use, you can easily delete and recreate the cluster on demand, thereby keeping costs low while still keeping your data persisted.

You can read more about Apache Hadoop at `https://hadoop.apache.org/`.

So, with this introduction and insight into Hadoop and Azure HDInsight, and with what you learned in Chapter 10 about Azure Data Lake Analytics, the question that was posed to you on the first page of this chapter should be ringing very loud. That question is, why would you choose to use Azure Data Lake Analytics over HDInsight and Hadoop, or vice versa?

Big Data Processing Decision Tree

The question of whether you should use HDInsight or Azure Data Lake Analytics really comes down to your needs, your skillset, and a few other items. Figure 12-1 should help provide some insight into the path you should choose.

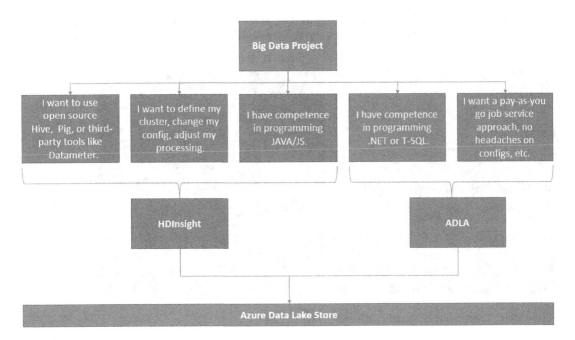

Figure 12-1. *Analytics decision tree*

In essence, if you want less of a "hands-off" approach and can sling C# and T-SQL code, Azure Data Lake Analytics should be considered. You learned about ADLA in Chapter 10, but I'll summarize here. ADLA lets you focus on gaining valuable insight from your data without worrying about clusters, hardware, or any type of infrastructure. Simply submit jobs and let ADLA do all the processing. ADLA provides dynamic scale, seamless integration, and is very cost effective via its pay-per-job model.

HDInsight, on the other hand, while still a full-managed service, is a bit more "hands on." Hadoop is a very complex system and it can be a bit difficult to manage, even in Azure because you still need to configure and adjust. There is a .NET SDK that allows you to create and run map/reduce jobs and execute Hive queries, but knowledge of JAVA still comes in handy.

Another thing to consider is if you have existing map/reduce code or if you are using Hadoop already, then the decision should be easy. But if you are coming into this new, Azure Data Lake Analytics is a strong solution for long-term storage with continuous data processing.

This section isn't to convince you or sway you one way or the other. It is simply to provide insight and guidance. Microsoft loves and supports both, so you'll get the same love and support regardless of which you choose. It all comes down to preference. If you decide to go with HDInsight, keep reading.

Creating a Cluster

Hopefully you kept reading simply because you want to learn about HDInsight. Great, because this section will walk you through the creation of an HDInsight cluster. This process has changed quite a bit since it was first in the portal, and to be honest, it will probably change some more. My hope is that it won't change so much that this section will be out of date.

Fire up your favorite browser and navigate to portal.azure.com and log in. When the portal dashboard appears, select New ➤ Intelligence + Analytics, and then select HDInsight, as shown in Figure 12-2.

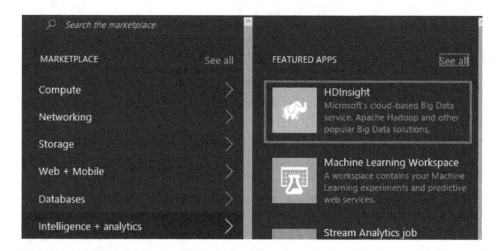

Figure 12-2. *Creating an HDInsight cluster*

The New HDInsight Cluster blade will appear which, at the time of this writing, looks like Figure 12-3. Obviously, there are a few things to fill out.

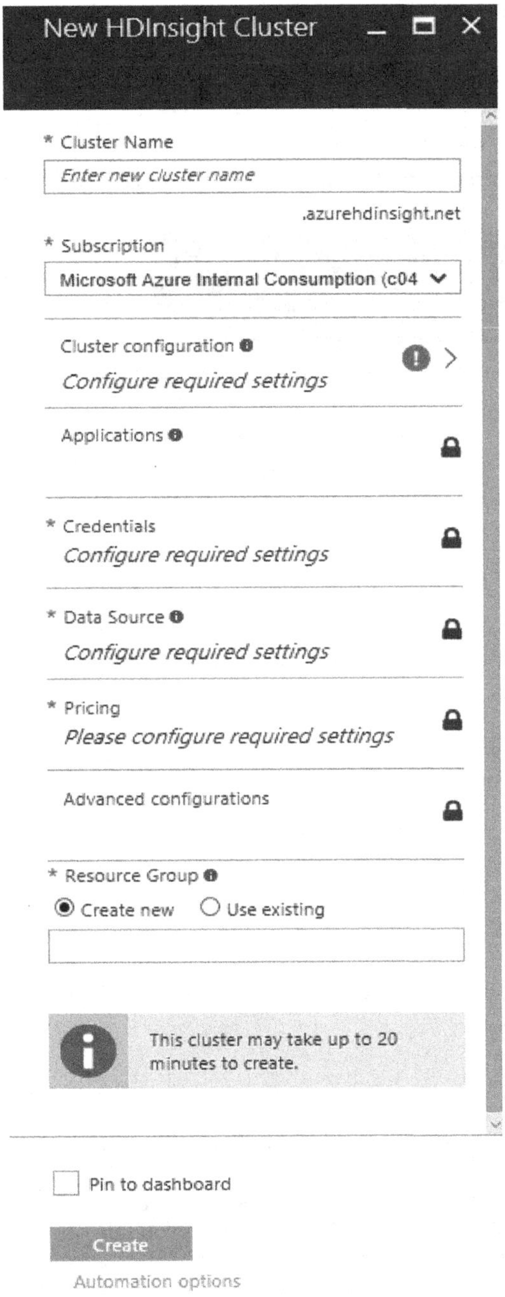

Figure 12-3. *New HDInsight cluster*

At the top of the New HDInsight Cluster blade you will see a message asking you to "click here" to try out the simpler, faster way of creating a cluster. Seriously, ignore it by clicking the Dismiss link. The "simpler" option creates a cluster with minimal configuration. That won't help you in this chapter.

Let's begin. First, enter a name for the cluster, make sure the proper subscription is selected, and then click in the Cluster Configuration area of the blade. This will open up the Cluster Configuration blade, shown in Figure 12-4.

Cluster Configuration

The Cluster Configuration blade is where you configure the cluster type, the operating system, version, and tier.

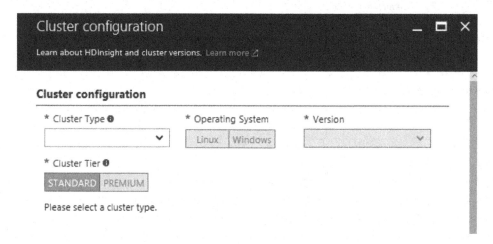

Figure 12-4. *HDInsight cluster configuration*

Currently there are seven types of HDInsight clusters you can create:

- Hadoop
- Hbase
- Storm
- Spark
- Interactive Hive
- R Server
- Kafka

At this current time, both Interactive Hive and Kafka are in preview so configuration settings and other aspects of these two cluster types may change.

For this example, select Hadoop as the cluster type, select the Linux operating system, select the latest version (which is Hadoop 2.7.3 as of this writing), and then select the Standard tier. As you select the configuration, you'll notice the available features, shown in Figure 12-5.

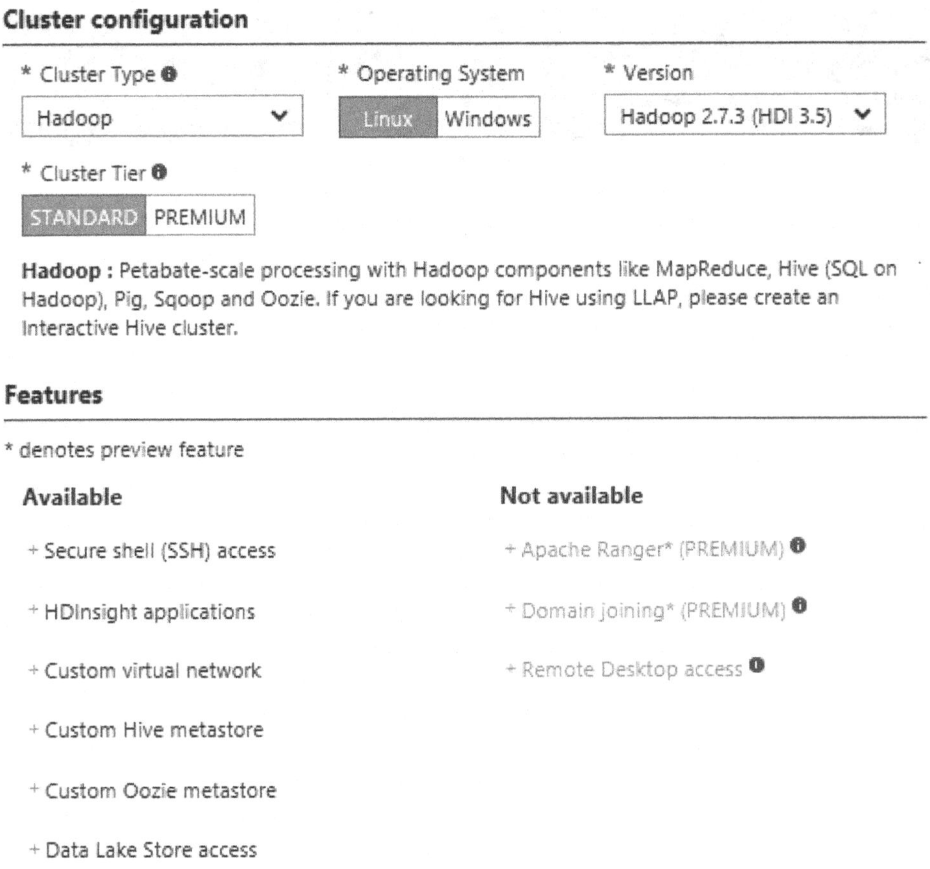

Cluster configuration

* Cluster Type ❶

| Hadoop ▾ |

* Operating System

| Linux | Windows |

* Version

| Hadoop 2.7.3 (HDI 3.5) ▾ |

* Cluster Tier ❶

| STANDARD | PREMIUM |

Hadoop : Petabate-scale processing with Hadoop components like MapReduce, Hive (SQL on Hadoop), Pig, Sqoop and Oozie. If you are looking for Hive using LLAP, please create an Interactive Hive cluster.

Features

* denotes preview feature

Available

+ Secure shell (SSH) access

+ HDInsight applications

+ Custom virtual network

+ Custom Hive metastore

+ Custom Oozie metastore

+ Data Lake Store access

Not available

+ Apache Ranger* (PREMIUM) ❶

+ Domain joining* (PREMIUM) ❶

+ Remote Desktop access ❶

Figure 12-5. *Cluster features*

Once the Cluster Configuration blade is filled out, click Select.

Credentials

The cluster credentials (the cluster login and password) on the Basics blade is where you specify the login and remote access for the cluster. The cluster login is automatically filled in with a username of admin, but you can change that if you'd like. Otherwise, simply enter a password of at least 10 characters in length and include one digit and one non-alphanumeric character. This username and password, as explained on the blade, are used for job submission and access to the cluster dashboard.

Next, enter an SSH username and password. This username and password are used to remotely connect to the cluster using a tool like PuTTY. Click Select on the Cluster Credentials blade.

Data Source

Next, click in the Data Source section of the New HDInsight Cluster blade, which will open up the Data Source blade. This is where things get a little tricky. The first part is easy, though. First, select Data Lake Store as the primary storage type, and then select your Azure Data Lake Store account that you created in Chapter 9.

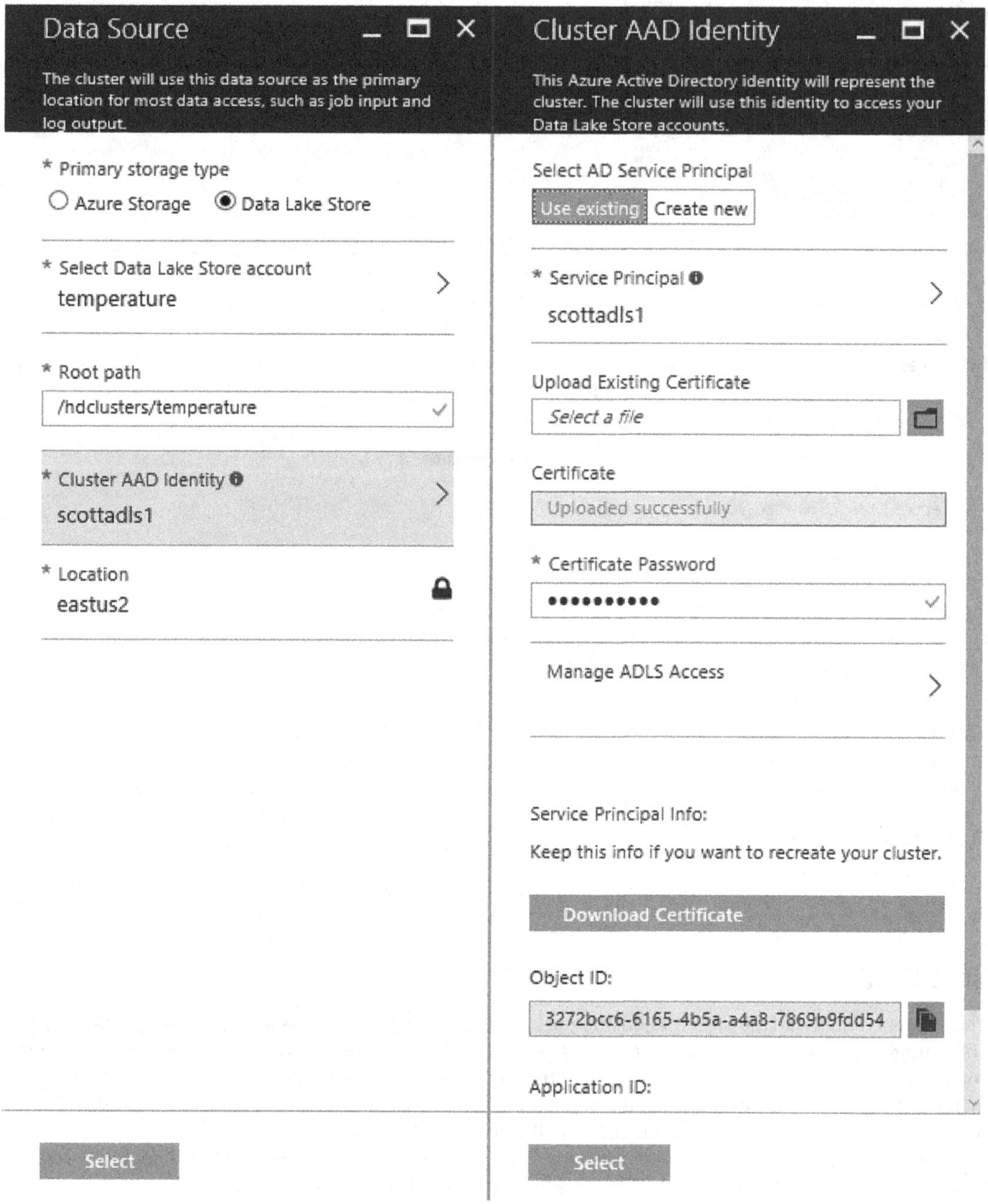

Data Source _ ▢ ✕

The cluster will use this data source as the primary location for most data access, such as job input and log output.

* Primary storage type

◯ Azure Storage   ⦿ Data Lake Store

* Select Data Lake Store account   〉
temperature

* Root path

| /hdclusters/temperature | ✓ |

* Cluster AAD Identity ❶   〉
scottadls1

* Location   🔒
eastus2

Cluster AAD Identity _ ▢ ✕

This Azure Active Directory identity will represent the cluster. The cluster will use this identity to access your Data Lake Store accounts.

Select AD Service Principal

| Use existing | Create new |

* Service Principal ❶   〉
scottadls1

Upload Existing Certificate

| Select a file | 🗀 |

Certificate

| Uploaded successfully |

* Certificate Password

| •••••••••• | ✓ |

Manage ADLS Access   〉

Service Principal Info:

Keep this info if you want to recreate your cluster.

| Download Certificate |

Object ID:

| 3272bcc6-6165-4b5a-a4a8-7869b9fdd54 | 📋 |

Application ID:

| Select |

| Select |

Figure 12-6. *Configuring the data source and service principle*

Next, in the Root path, enter the Azure Data Lake Store path where the data you want to process exists. For this example, I went into Azure Data Lake Store and created a directory called `hdclusters` and then I created a subdirectory called `temperature`. This is simply for the purpose of this example. In the examples throughout this book and especially in Chapter 6, my temperature data from my sensors goes into the \ `tempcluster\logs` directory prefixed by the date in Azure Data Lake Store. To make the example in this chapter a bit easier, I created a simple directory structure. You can see this in Figure 12-6. I also used this special folder because the HDInsight creation will put its files here, similar to what it does with Azure Blob Storage. Thus, I wanted to differentiate it from the location where my Azure Steam Analytics drops the sensor data.

OK, back to the example. Next, click the Cluster AAD Identity section in the Data Source blade. This will open up the Cluster AAD Identity blade, shown in Figure 12-6. This is where things get a bit tricky. Select "Create new" for the Select AD Service Principal and then click the Service Principal section, which will open the Create Service Principal blade.

In the Create Service Principal blade, enter a Service Principal name, provide a Certificate expiration date, and provide a certificate password (confirm the password). Click Create on the Create Service Principal blade.

At this moment, the service principal with the name you provided will automatically be created as soon as you click Create. The Create Service Principal blade will close and you'll be taken back to the Cluster AAD Identity blade. At this point, the Download Certificate button will be enabled and the Object ID and Application ID will automatically be filled in on the Cluster AAD Identity blade. Before you click Select on the Cluster AAD Identity blade, download the certificate. You'll thank me later.

Now you have two options. Even though the Cluster AAD Identity blade is filled out, if you click Select on the Cluster AAD Identity blade, you will get a red exclamation mark in the Cluster AAD Identity section of the Data Source blade. Why? Because if the service principal has been created, that service principal needs rights to the directory you specified in the Root path on the Data Source blade.

Therefore, before you continue, you need to go into your Azure Data Lake Store account and give your service principal specific rights to the directory. To do that, your two options are

- Cancel out of the cluster creation altogether and go into your Azure Data Lake Store account. Yes, you'll need to start the cluster creation all over.

- Open up a second browser tab, open the Azure portal, and go into your Azure Data Lake Store account.

Seriously, if you have to think about this, we need to talk. At some point engineering will fix this, but for now, these are your two solutions, and honestly, you had better pick the second one.

So, once you are in the portal in your second tab, navigate to your Azure Data Lake Store account and click the Data Explorer link. Navigate to the directory you created where your data sits and which you specified in the Root path.

First, navigate to the subfolder (in this example, it's the `temperature` folder). Once there, click the Access button. In the Access blade, click Add. In the Select User or Group blade, type the name of the service principal you created during the cluster creation process. The list will automatically filter as you type. Once you find your service principal, select it and click Select. The Select Permissions blade will automatically open.

Pay particular attention to these next steps as they are best practice and will make that red exclamation point go away. In the temperature subdirectory, give your service principal Read, Write, and Execute permissions. Do not change anything else. Then click OK.

Next, in the Data Explorer, navigate up a folder (in this example, it's the hdclusters folder). Repeat the same process for adding your service principal to this folder as you did for the subfolder. Grant Read, Write, and Execute permissions, don't change anything else, and then click OK.

Next, go back up to the root folder, which is the name of your Azure Data Lake Store account, and repeat the process EXCEPT this time only grant the service principal Execute permissions, nothing else.

With all of that completed, go back to your first tab, click Select on the Cluster AAD Identity blade, and all should be swell.

Pricing

Lastly is the Pricing blade. By default, the number of worker nodes will default to 4. You can keep that or change it. For this example, I don't need that many so I changed it to 2. Next, click the Worker node size section of the Pricing blade, which will open up the Choose Your Node Size blade.

By default, the Choose Your Node Size blade will only show 3 sizes. To view more, click the View All link in the right-hand corner of the blade. Notice that you have, as of this writing, 15 sizes to choose from, ranging from a general purpose A3 with 4 cores and 7GB of RAM to an optimized D14 with 16 cores and 112GB RAM, 32 disks and 800GB of local SSD storage and 35% faster CPU. Yeah, that's cool.

For this example, I picked middle-of-the-pack D3s with 4 cores and 14GB of RAM. Pick whatever you feel you need and then click Select. Repeat the process for the Head node. Notice that the Pricing blade will give you an estimated total cost per hour of running this cluster, as shown in Figure 12-7. This is helpful when planning what your overall cost will be per month for your solution.

Click Select on the Pricing blade.

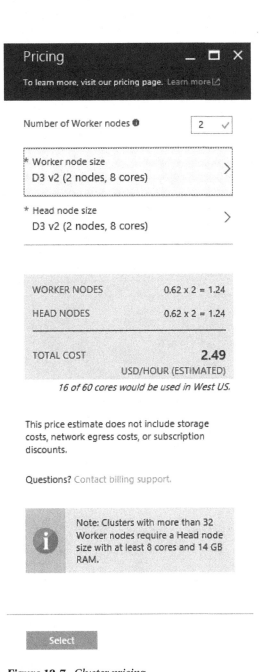

Figure 12-7. *Cluster pricing*

At this point, you should be back on the New HDInsight Cluster blade with it all filled out and it should look something like Figure 12-8. Be sure to select "Pin to dashboard box" and then click Create. As it states on the New HDInsight Cluster blade, it could take up to 20 minutes to create the cluster. I've never had it take that long, but it depends on how many nodes you are creating.

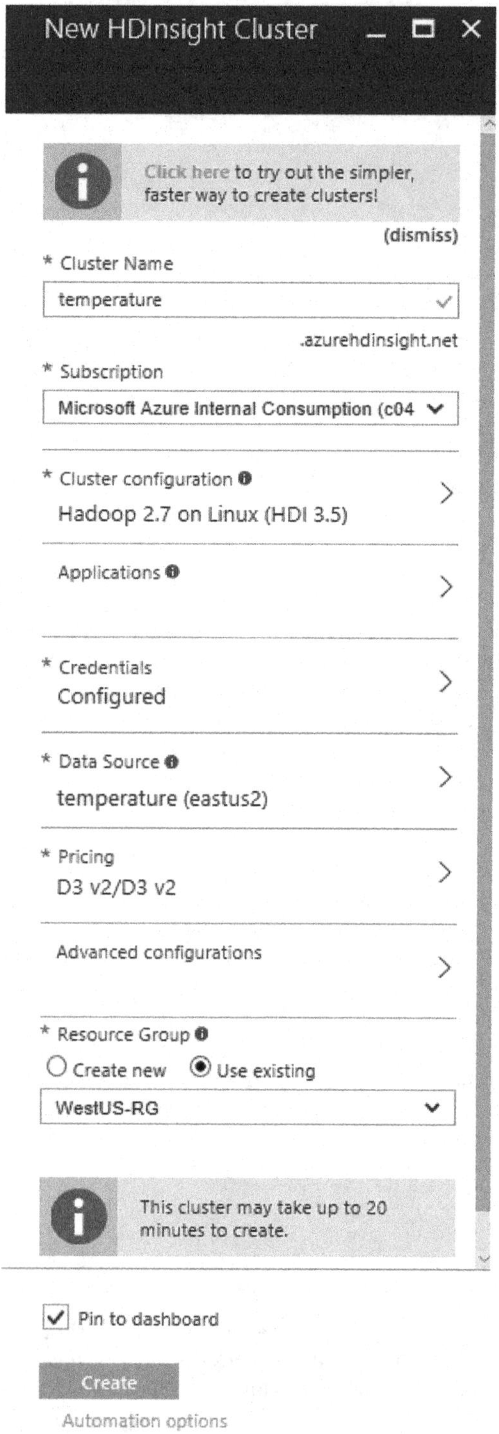

Figure 12-8. *Finished HDInsight cluster configuration*

Congratulations on creating your HDInsight cluster. Sit back and relax while your cluster is being provisioned; you deserve it. You're going to put it to good use in the following section.

Using Hadoop for Batch Queries

This section will discuss several ways to use Hadoop in Azure (HDInsight) to execute batch queries with Hive. There are many technologies available to submit Hive queries, including the following:

- SSH
- Beeline
- cURL
- PowerShell
- Java UDF

However, the following sections will use the .NET SDK, HDInsight Tools for Visual Studio, and the Query Console. Documentation for other methods can be found online in the How To section at https://docs.microsoft.com/en-us/azure/hdinsight/.

.NET SDK

With the HDInsight cluster created, the following three sections will show you how to submit Hive queries using a variety of methods, beginning with this section, which will show you how to use the .NET SDK.

Fire up Visual Studio 2015 and create a new C# console application and name the project whatever you like. Once the project is created, you need to install the HDInsight Job library from Nuget, which contains all the classes to submit jobs.

Thus, from the Nuget Packager Manager Console, type the following command, as shown in Figure 12-9:

```
Install-Package Microsoft.Azure.Management.NDInsight.Job
```

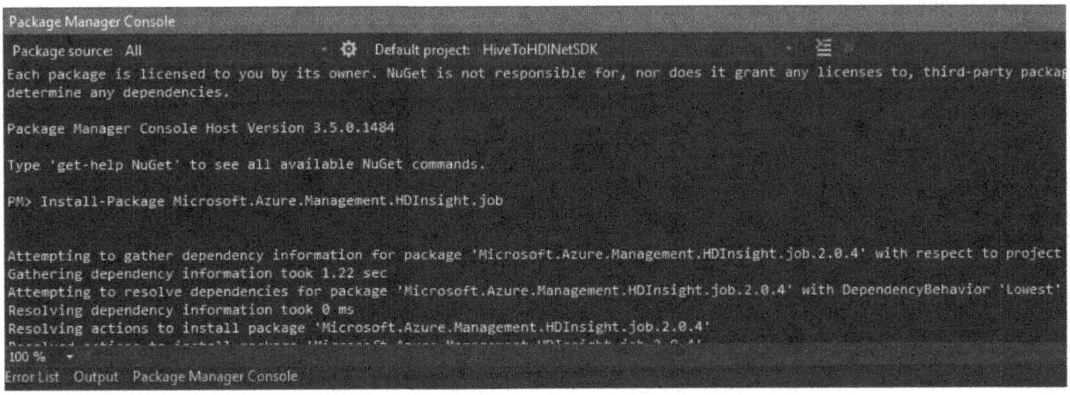

Figure 12-9. *Installing the Microsoft Azure HDInsight Job Management Library*

It will only take a minute or less to install the package, but once complete, open Solution Explorer and expand the References node. Notice that this package installed a lot of dependency assemblies including Newtonsoft.Json, Microsoft.Data.Odata, and even one called Hyak.

Hyak provides the framework for common error handling, tracking, configuration, and HTTP/Rest-based manipulation for any clients making REST calls. Thus, it makes sense to include it.

Next, add the following using statements to `Program.cs`:

```
using System.Threading;
using Microsoft.Azure.Management.HDInsight.Job;
using Microsoft.Azure.Management.HDInsight.Job.Models;
using Hyak.Common;
```

Next, add the following to `Program.cs` directly above the `Main()` method. These statements declare the variables that contain the cluster information to be able to submit jobs.

```
private static HDInsightJobManagementClient _hdiJobManagementClient;
private const string ExistingClusterName = "<Your HDInsight Cluster Name>";
private const string ExistingClusterUri = ExistingClusterName + ".azurehdinsight.net";
private const string ExistingClusterUsername = "<Cluster Username>";
private const string ExistingClusterPassword = "<Cluster User Password>";
```

The next step is to add the code that will authenticate the user submitting the job, create an instance of the job management client, and then call a method to create and submit a job. Thus, add the following code to the `Main` method:

```
System.Console.WriteLine("The application is running ...");

var clusterCredentials = new BasicAuthenticationCloudCredentials { Username =
ExistingClusterUsername, Password = ExistingClusterPassword };
_hdiJobManagementClient = new HDInsightJobManagementClient(ExistingClusterUri,
clusterCredentials);

SubmitHiveJob();

System.Console.WriteLine("Press ENTER to continue ...");
System.Console.ReadLine();
```

The last step is to create the method and code to submit the Hive job. Add the following code below the `Main` method. This is the `SubmitHiveJob` method which is called from the `Main` method. The code in the `SubmitHiveJob` method creates a Hive job, defines the Hive query including the query parameters and arguments, and submits the query.

```
private static void SubmitHiveJob()
{
    Dictionary<string, string> defines = new Dictionary<string, string> { { "hive.execution.
engine", "ravi" }, { "hive.exec.reducers.max", "1" } };
    List<string> args = new List<string> { { "argA" }, { "argB" } };
    var parameters = new HiveJobSubmissionParameters
    {
```

```
        Query = "CREATE EXTERNAL TABLE StationUsage (Station string, Gate_location string,
TranDate string, Day_type int, Flag int, Entry_or_exit string, Total_day_count int) ROW
FORMAT DELIMITED FIELDS TERMINATED BY ',' STORED AS TEXTFILE;",
        Defines = defines,
        Arguments = args
    };

    System.Console.WriteLine("Submitting the Hive job to the cluster...");
    var jobResponse = _hdiJobManagementClient.JobManagement.SubmitHiveJob(parameters);
    var jobId = jobResponse.JobSubmissionJsonResponse.Id;
    System.Console.WriteLine("Response status code is " + jobResponse.StatusCode);
    System.Console.WriteLine("JobId is " + jobId);

    System.Console.WriteLine("Waiting for the job completion ...");

    // Wait for job completion
    var jobDetail = _hdiJobManagementClient.JobManagement.GetJob(jobId).JobDetail;
    while (!jobDetail.Status.JobComplete)
    {
        Thread.Sleep(1000);
        jobDetail = _hdiJobManagementClient.JobManagement.GetJob(jobId).JobDetail;
    }
}
```

When the job is submitted, a response and JobId are sent back. Technically, a JobId is only sent back if the job is submitted and accepted. Notice in Figure 12-10 that the status for the job submitted is "OK" and thus a JobId is returned.

The last bit of code simply loops, checking for job status. If the status is not "complete," it waits for a bit and then checks again. When the job is completed, the results of the query are returned. However, in the screenshot of Figure 12-10, I did not include the results because an example below will show the results as you explore a different way to submit Hive queries.

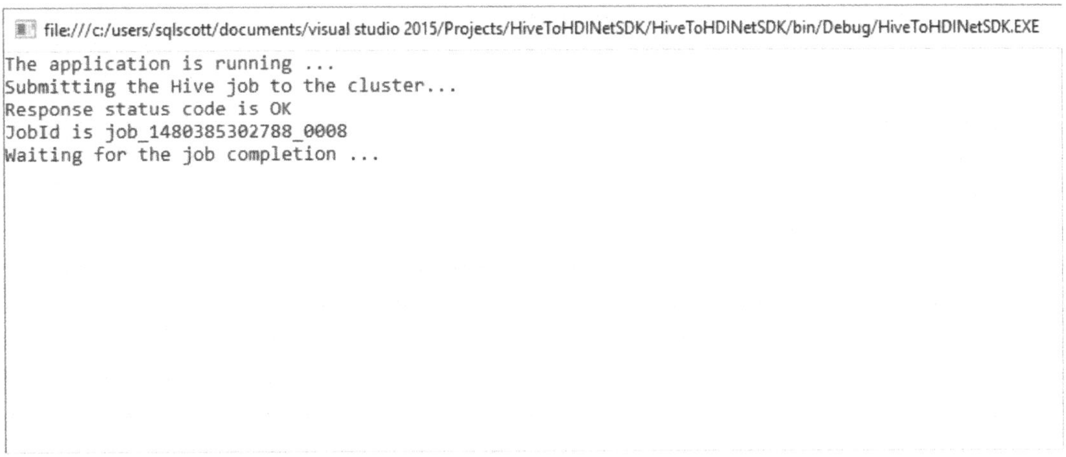

Figure 12-10. *Submitting the job to the HDInsight cluster*

The query used in this example is the following:

```
CREATE EXTERNAL TABLE StationUsage (Station string, Gate_location string, TranDate string,
Day_type int, Flag int, Entry_or_exit string, Total_day_count int) ROW FORMAT DELIMITED
FIELDS TERMINATED BY ',' STORED AS TEXTFILE;
```

The query creates a new external table in Hive. External tables only store the table definition in Hive and the data is left in the original location. Since your data is stored in Azure Data Lake Store, you'd like to leave it there and only apply the schema of the table upon read (when a SELECT statement is executed).

The takeaway from this simple example is that it is very easy to create and submit Hive queries using the .NET SDK.

HDInsight Tools for Visual Studio

This example also uses Visual Studio but uses HDInsight Tools. At the time of this writing, you can download the tools from www.microsoft.com/en-us/download/details.aspx?id=49504. Click the Download button and download the appropriate file for the version of Visual Studio you are using (2013 or 2015). Save the file, and once downloaded, run the installer. The install is straightforward: select the defaults through the installation wizard and click Install. After a minute or so the install will complete and you'll have everything you need to work with Azure Data Lake, including HDInsight.

The next thing to do is to expand the Server Explorer window and connect to your Azure subscription. You should see an Azure node in the Server Explorer window. Right-click the Azure node and select "Connect to Microsoft Azure Subscription" from the context menu. Enter your Azure subscription credentials in the pop-up and then click "Sign in." You'll only need to do this once per workstation. Right-click the Azure node again and select Refresh from the context menu, and then expand the HDInsight node; you should see your HDInsight cluster in the list.

Create a new project, and in the list of templates you will see a new node called Azure Data Lake. Expand this node, click the Hive node, and select the Hive Application template, as shown in Figure 12-11.

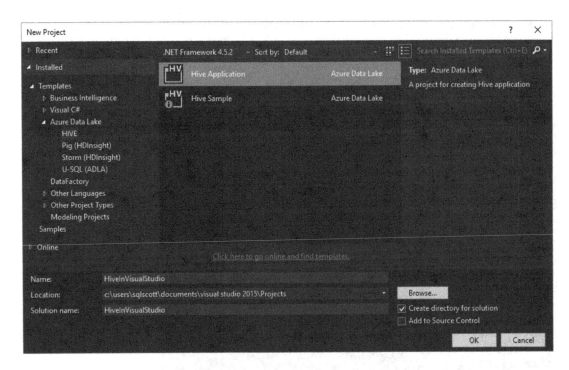

Figure 12-11. *Create a new Hive application in Visual Studio*

Once the project is created, a new Hive script is also generated and opened. In the Hive query window, type in the following:

```
set hive.execution.engine = tez;
DROP TABLE StationUsage;
CREATE EXTERNAL TABLE StationUsage (Station string, Gate_location string, TranDate string,
Day_type int, Flag int, Entry_or_exit string, Total_day_count int)
ROW FORMAT DELIMITED FIELDS TERMINATED BY ','
STORED AS TEXTFILE;
SHOW TABLES;
```

This code will drop the `StationUsage` table if it exists and then create it. This is the same table created in the first example. However, it also does two other things. First, it sets the Hive execution engine to Tez. Tez is an extensible framework that improves MapReduce performance. Second, the query executes `SHOW TABLES`, which is the same thing as `SELECT * FROM sys.Tables` in SQL Server. Your query window should now look like Figure 12-12.

Figure 12-12. *Hive query window*

To run the query, simply click the Submit button on the query window toolbar. The Hive job summary appears and displays information about the running job. Click the Job Output link to see the results of the job, as shown in Figure 12-13.

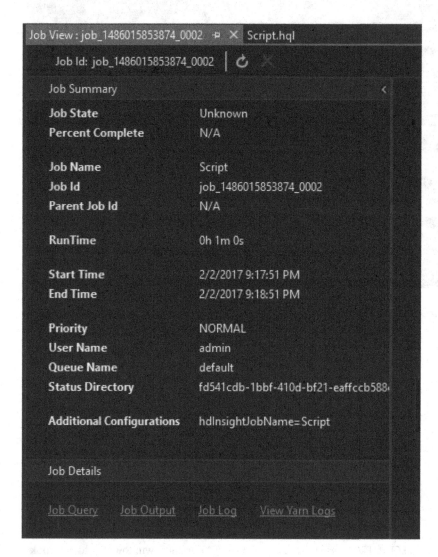

Figure 12-13. Viewing the Hive job output

To create another query script, you can right-click the solution, select Add ➤ New Item, and select the Hive Script item. An easier way is to right-click the HDInsight node in Server Explorer and select "Write a Hive Query" from the context menu, as shown in Figure 12-14. You also have the ability to view currently running jobs by selecting View Jobs from the context menu.

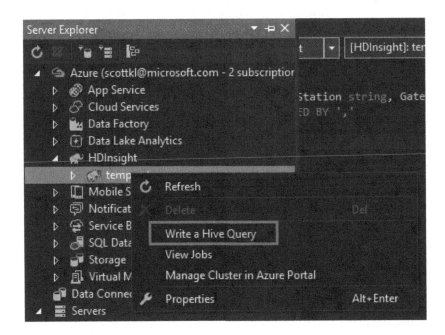

Figure 12-14. *Running a Hive query from Server Explorer*

The context menu also offers the ability to view currently running jobs.

Query Console

This last example will show you how to use the query console to run Hive jobs. The last few examples have required Visual Studio, which is great when you need to run Hive jobs in production and have more control over the environment. However, if you simply want to test a Hive query, there is no better place, in my opinion, than the query console.

When the cluster was created, the example used the Linux operating system, and in order to manage, monitor, troubleshoot, and run queries you need to use Ambari. Apache Ambari is a project aimed at making Hadoop management easy. For HDInsight, you open the Ambari Dashboard via the Azure Portal. Open the HDInsight cluster and right in the middle of the Properties page is the Ambari Views link, shown in Figure 12-15.

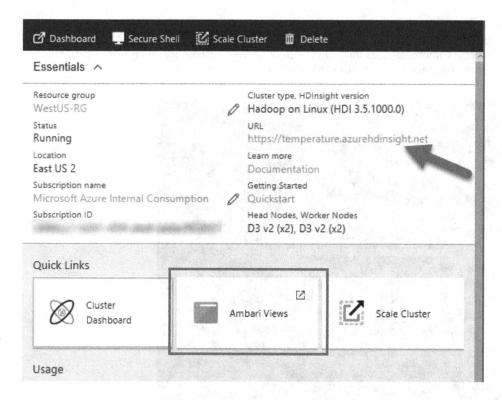

Figure 12-15. *Opening the Ambari view*

Clicking the Ambari Views link opens the Ambari View. When the page opens, click the Dashboard link on the menu bar, as shown in Figure 12-16. This dashboard gives you a nice glance at what is happening in your Hadoop cluster. Within this dashboard, you can monitor the health and overall status of the cluster, gather metrics, or set up alerting for all sorts of issues such as when a node goes down, low disk space, and more.

To run Hive queries, click the icon next to the Admin menu, and it will drop down a menu that includes an option called Hive View, as shown in Figure 12-16. Select that option.

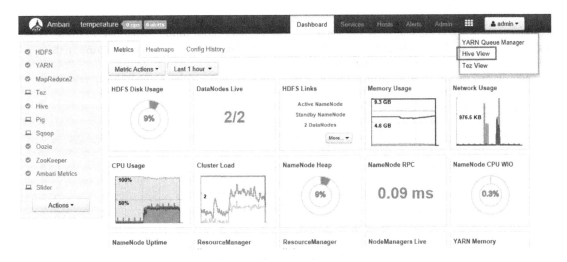

Figure 12-16. *The Ambari dashboard*

Taking a look at the Hive Query page in Figure 12-17, you should notice several things. On the left is the Database Explorer. Right now, there is only the default database but, if needed, additional databases could be created. On the right is a list of icons that will allow you to work with HiveQL. The two you will use in this example are the top one, SQL, and the bottom one with a message icon. The middle section is where queries are written and executed.

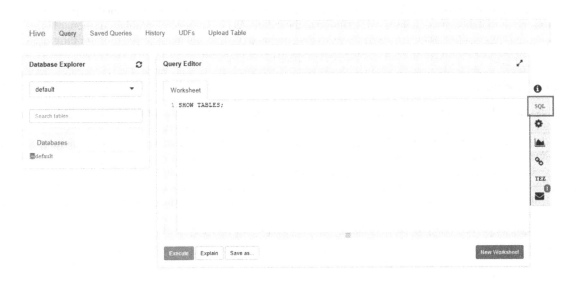

Figure 12-17. *The Hive Query Editor*

For this example, you'll execute a simple query. Simply type in SHOW TABLES;. As explained earlier, this is similar to the T-SQL statement SELECT * FROM sys.tables. Well, more appropriately, SELECT name FROM sys.tables. Click the Execute button and you will immediately notice a new message appearing on the Message icon, as well as some text stating that the query has been submitted successfully.

After several seconds, the text will change to Status: SUCCEEDED and the results of the query will display on the Results table below the query, as shown in Figure 12-18.

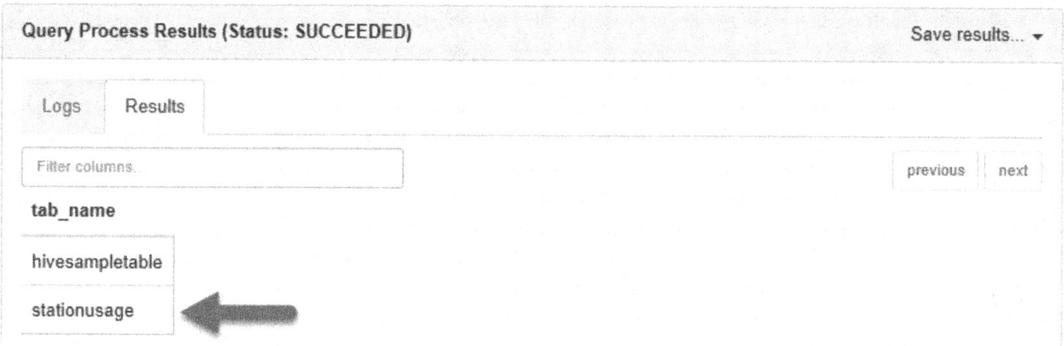

Figure 12-18. *Checking the results*

In the results you will notice that the table StationUsage exists, which was created from the first example earlier. If you ran the second example, the table was then dropped and recreated.

Before wrapping up, click the Notification icon. This page shows queries submitted, success and failure messages, and if there was a failure, a description of the failure. As I mentioned, the Ambari dashboard is a quick and easy way to execute a query, look at errors generated (if any), and manage and monitor the cluster.

Summary

This chapter focused on Azure HDInsight, a solution for Hadoop-based big data batch processing. As big data processing and analytics comes into play within an organization, the need to efficiently gain insights into that data becomes necessary. HDInsight provides this capability in the cloud by making provisioning, managing, and maintaining a big data cluster easy.

In this chapter, you got an overview of Hadoop and thus HDInsight, and then you explored the differences between HDInsight and Azure Data Lake Analytics. With that foundation, you created an HDInsight cluster and looked at the different technologies and methods for writing and executing batch queries.

Chapter 13 will focus on real-time data insights and reporting with Power BI.

CHAPTER 13

■ ■ ■

Real-Time Insights and Reporting on Big Data

In the early chapters of this book, I talked about hot path and cold path avenues for data. Up until now, the data has followed a cold path in that the data was picked up by Azure Stream Analytics and routed to storage for downstream processing. The data stores were Azure Blob Storage and Azure Data Lake Store, and they provided a means for analytic processing via HDInsight and other analytic engines and services. For example, the last chapter talked about processing the data using Azure HDInsight, using Azure Data Lake Store as the data source with data that Azure Stream Analytics had dropped in there.

This chapter will focus on the hot path scenario, which provides real-time streaming and analysis to incoming data. Whereas cold path deals with batch processing on data, a key value of hot path data is that it allows users to gain real-time insights into your data.

As discussed in earlier chapters, the data can come from a myriad of sources including devices, sensors, and applications. Real-time data insights are good for connected IoT scenarios, such as the connected car and connected home, and as well as health and farm scenarios. The example in this book uses data generated from temperature sensors hooked up to a Raspberry Pi and a Tessel.

What you have learned so far is that Azure Stream Analytics together with Azure IoT Hub offer the capabilities to stream millions of events per second to provide the much-needed analytics in real time. However, what is missing, or what has not been discussed so far, is the mechanism for visualizing the real-time analytics. This chapter will discuss exactly that.

Real-Time Streaming and Analytics with Power BI

Chapter 6 introduced Azure Stream Analytics, which provides real-time data stream processing in the cloud. In subsequent chapters, Azure Stream Analytics was used to stream data to data stores for cold path batch processing using HDInsight and Azure Data Lake Analytics.

As you learned in Chapter 6, a streaming analytics service can be utilized for a variety of use cases, such as analyzing IoT data, social media sentiment analysis, fraud detection, and real-time sporting and gaming statistics.

However, as part of your hot path scenario, you want to quickly and efficiently analyze the data that is coming in to see what is happening in real time. To do that, you need a tool that provides rich reporting through interactive dashboards and reports with tools to transform and visualize the incoming data. To do that, this chapter will discuss and use Power BI. Granted, there are other tools available, such as Tableau, which can be used to view or connect to data and these tools do a fantastic job, but Azure Stream Analytics and Power BI provide a seamless integration for real-time analytics.

© Scott Klein 2017
S. Klein, *IoT Solutions in Microsoft's Azure IoT Suite*, DOI 10.1007/978-1-4842-2143-3_13

This chapter will not be an in-depth discussion on Power BI, but rather it will show how to hook up Power BI to Azure Stream Analytics to stream data and update Power BI dashboard in real time. This chapter will assume that you have a Power BI account. If you do not, it is easy to get started. Simply open your favorite browser and navigate to www.powerbi.com. Smack dab in the middle of the page near the top is a button that says "Get started free," as shown in Figure 13-1. Click it.

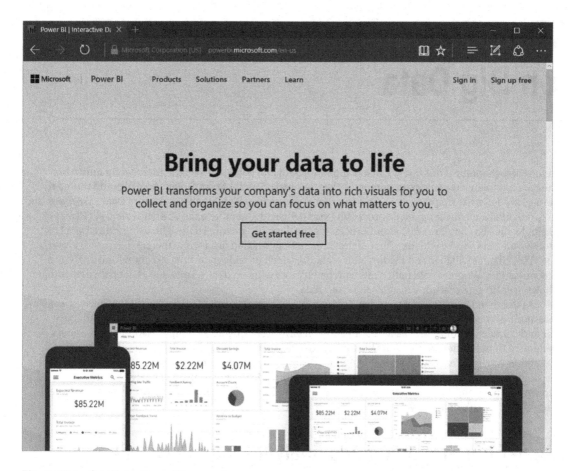

Figure 13-1. *Creating a Power BI account*

You will be presented with two options; Power BI Desktop or Power BI service. I won't go into too much detail here, but Power BI Desktop is simply a Power BI desktop application that lets you connect to data and create visual reports. These reports can then later be shared by uploading them to the Power BI service.

This example will be using the other option, the Power BI service, which is URL-based. Click the Sign up button and walk through the sign up process to create a Power BI account. Once done, you will be ready to start connecting to data sources and writing reports. However, you first need to configure Azure Stream Analytics to send data to Power BI.

Configuring Azure Stream Analytics

I'll get to the Power BI piece shortly, but the first piece to setting this up is to configure Azure Stream Analytics to send the data to Power BI. If you haven't fully grasped the full potential of what was just said, put the book down and think about this for a bit. Power BI integrates with Azure Stream Analytics, meaning that you can now output the results of a stream analytics job directly to Power BI where it can be visualized, explored, and even shared via a real-time dashboard.

It is this technology that helped a farm build a cow-monitoring system to give the farmer real-time insights into his cow herd, thereby boosting milk production, increasing cow production, and ensuring healthy cows.

Awesome, isn't it? And, it's actually quite easy. Open your favorite browser and navigate to the Azure portal page at `portal.azure.com`. Once logged in, open the Stream Analytics job that you have been using in the demo throughout this book.

Just like Azure Blob Storage and Azure Data Lake Analytics, you will need to add an output to Power BI. If your Stream Analytics job is still running, you will need to stop it in order to add the new output.

In the Job Topology on the Essentials page, click the Outputs tile to open the Outputs blade. In the Outputs blade, click Add, which will open the New Output blade, shown in Figure 13-2. In the New Output blade, give the output alias a unique name and then select Power BI.

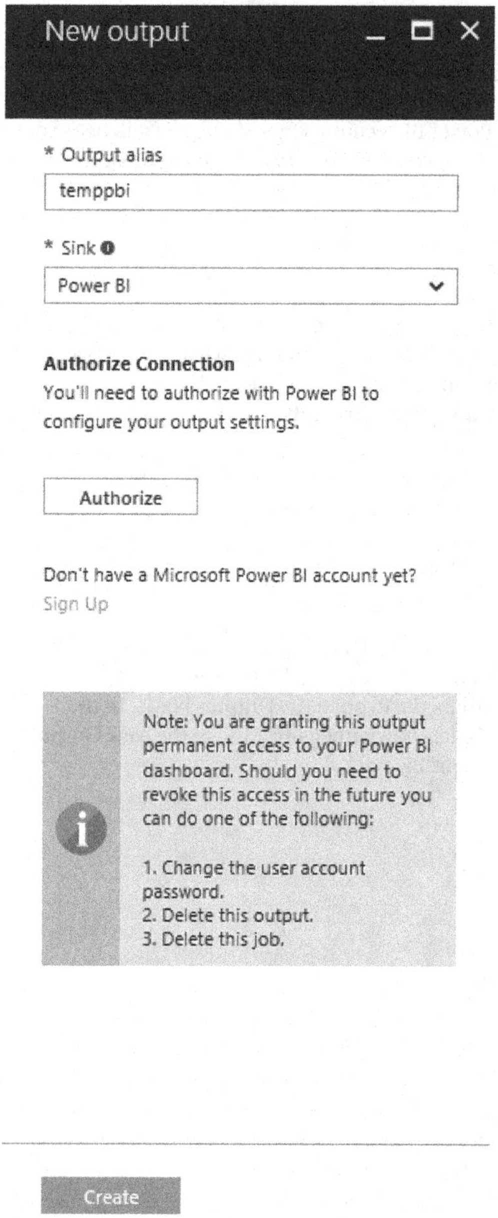

Figure 13-2. *Selecting Power BI in the New Output blade*

You will notice an Authorize button on the Output blade with a message stating what will be happening. Essentially, you need to give Azure Stream Analytics permission to access your Power BI account. Thus, go ahead and click Authorize, which will pop up a dialog and ask you to enter the same credentials you used to create your Power BI account.

Once the authorization is complete, you will be asked to select a Group Workspace and provide a dataset name and table name. Groups in Power BI are a way to implement a level of security and manage workspaces. Each workspace can contain its own datasets, reports, and dashboards. When you create your account, you will notice that there is a default My Workspace. It will suffice for this example, so select it and provide a dataset name and table name, as shown in Figure 13-3.

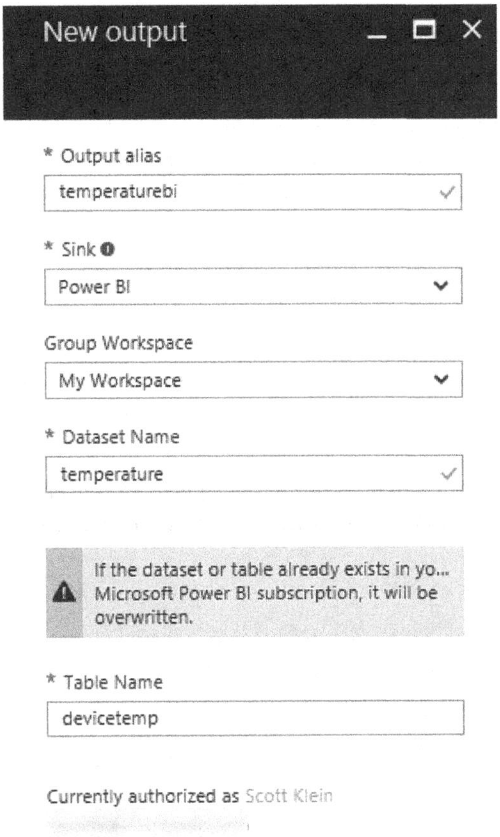

Figure 13-3. *Configuring the Azure Stream Analytics output*

If you were to go to Power BI now, you would not see the dataset in the list of datasets yet, as shown in Figure 13-4, simply because it is created during runtime, meaning that when the Azure Stream Analytics job runs, it will create the dataset at that time.

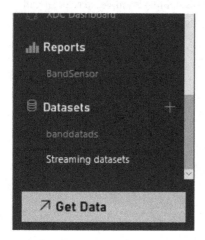

Figure 13-4. *Listing the datasets*

Click Create on the New Output blade to add the Power BI output to the list of available outputs in your Azure Stream Analytics job. Close the Outputs blade and you should now see three outputs for the Stream Analytics job, as shown in Figure 13-5.

Figure 13-5. *The three outputs of the Azure Stream Analytics job*

With the output created, the next step is to modify the query to send the data to the Power BI output. Click the query tile and add the query shown in the red section of Figure 13-6.

Need help with your query? Check out some of the most common

```
1  SELECT
2      *
3  INTO
4      tempblogstorage
5  FROM
6      iothubinput
7
8  SELECT
9      *
10 INTO
11     temperatureadls
12 FROM
13     iothubinput
14
15 SELECT
16     MAX(Temp) AS temp,
17     MAX(Humidity) AS hmdt,
18     System.TimeStamp AS time,
19     Device AS device,
20     Sensor AS sensor
21 INTO
22     temperaturebi
23 FROM
24     iothubinput TIMESTAMP BY time
25 GROUP BY
26     TUMBLINGWINDOW(ss,1),
27     device,
28     sensor
```

Figure 13-6. *Modifying the Azure Stream Analytics query*

You'll notice an immediate error message stating the following:

"Input source with name iothubinput' is used in both with and without TIMESTAMP BY clause. Please either remove TIMESTAMP BY clause or update all references to use it."

What this message is stating is that you have three queries in the query window and only one of the queries uses the TIMESTAMP BY clause, and either all three need to use the clause or none of them. It's an all-or-nothing message.

At this point, there are two options: remove the other two queries or modify the other two queries to include the TIMESTAMP BY clause. For the purposes of this demo, I just removed the other two queries. I also modified the query slightly to remove the sensor column from the GROUP BY. I only really need to group by the device, not both the device and sensor.

```
SELECT
    MAX(Temp) AS temp,
    MAX(Humidity) AS hmdt,
    System.TimeStamp AS time,
    Device AS device
INTO
    temperaturebi
FROM
    iothubinput TIMESTAMP BY time
GROUP BY
    TUMBLINGWINDOW(ss,1),
    device
```

Save the query and then click the Start button on the Stream Analytics job, as shown in Figure 13-7.

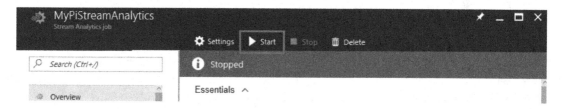

Figure 13-7. *Starting the Stream Analytics job*

At this point, you are done with the Azure side of things. Even now if you were to go to Power BI, you would not see the dataset in the list of datasets yet because no data is being sent. It will be created at runtime when there is data being sent. Thus, you need to start sending data. So, fire up your Raspberry Pis and Tessels, and start sending temperature data.

In this example, I fired up a Raspberry Pi with a DHT22 sensor and let it run for a minute. I then took an air canister and hair dryer and rotated blowing hot air and cold air on it for a few minutes. You will see the outcome shortly. If you are following along, you will want to do the same thing, meaning find some way to fluctuate the temperate.

Before we move on to the Power BI piece, a few more words about the query that was used: notice that it uses the TIMESTAMP BY in the FROM clause as well as a TUMBLINGWINDOW in the GROUP BY clause.

All data stream events have a timestamp associated with them and by default they are timestamped on their arrival time in the input source. To ensure that the data is in the order of when it was generated, the TIMESTAMP BY clause allows you to specify the time from the data being sent and not the time of when the data arrived. Now, for the purpose of this example, you are using a system-generated time so you don't need to include it in the GROUP BY clause. Not critical for this example, but if you were to use the time from your application, the query would complain and make you include it in the GROUP BY. So let's leave it out for now.

Next, notice the TUMBLINGWINDOW. We discussed this in Chapter 6 but I'll revisit here for the purposes of this demo. A tumbling window is a series of fixed-size, non-overlapping, and contiguous time intervals. In this case, I am sending the data to Power BI in one-second intervals because I want Power BI to update in real time every second. The TUMBLINGWINDOW gives me the contiguous time intervals that don't overlap, so I'll get the data as one-second streams into Power BI. OK, now on to Power BI.

Power BI

Go back to your Power BI browser window. If everything is working correctly, meaning the devices are sending data and Azure Stream Analytics is picking the data up from Azure IoT Hub and sending it to Power BI, you should see the `temperature` dataset appear in the list of available datasets in Power BI, as shown in Figure 13-8.

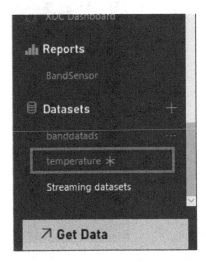

Figure 13-8. *List of available datasets*

So now you know that data is being sent by Azure Stream Analytics and received by Power BI. It is now time to build the real-time dashboard.

Creating the Report

I like how easy it is to get started with Power BI. I am certainly not a Power BI guru by any means, but the fact that I can get around in Power BI and build a dashboard certainly attests to the usability and friendliness of Power BI. Honestly, the hardest part of figuring out how to get the report to display the data as I wanted it was not in the building of the report but in the Stream Analytics query. Once I had the query figured out, the report took no time at all.

Trust me, if you want real-time analytics, spend some time learning how the tumbling window, hopping window, and sliding window statements work. Experiment with them to see how they affect your query and its output. My homework assignment, and your homework assignment, is to go back to the query in Azure Stream Analytics and add the two queries back that send data to Azure Blog Storage and Azure Data Lake Store using a similar query and see what results you get.

OK, on to building the report. Simply clicking on the temperature dataset presents you with a blank canvas in which to build your report, as shown in Figure 13-9. Here is where you will add the graphs, charts, and whatever else to display the real-time temperature data.

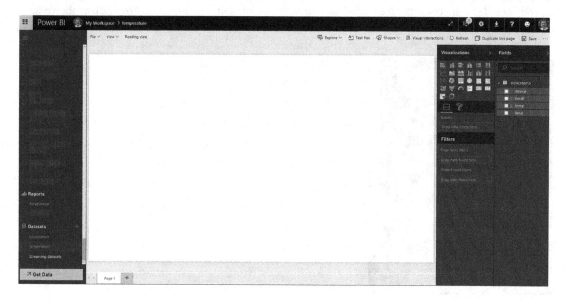

Figure 13-9. *Creating the report*

To the right of the blank canvas is the list of visualizations and fields in which to build and include in your report. As shown in Figure 13-10, the visualizations include different graphs and charts, along with a list of fields found in your dataset. These fields are then used determine what data and values from your dataset are to be used in the report as well as filter the report data.

To build the report, select the Line Chart report type, which is in the far left column, second row. Selecting that type of report will place a generic line chart on your report. You want the temperature displayed by time along the X axis, so drag the time column to the Axis field. You want to display the temperature, obviously, so drag the temp column to the Values field. You want the legend on the graphs to show the device name so you can correlate temperatures to devices, so drag the device column to the Legend field.

To make it easy, you'll have one device per graph, so let's filter this line chart on a specific device. Drag the device column down to the Visual Level Filters field and select a device from the list of available values. When you are all finished with your configuration it should look like Figure 13-10.

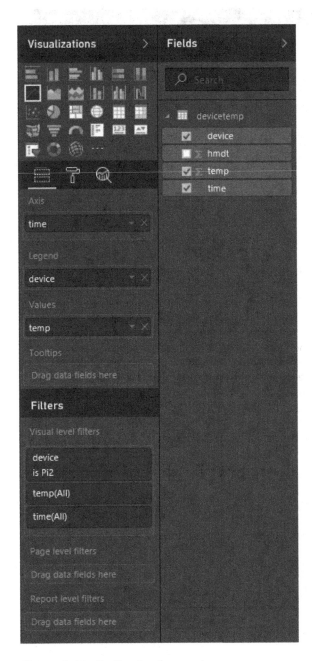

Figure 13-10. *Configuring the report*

Your report should now look like Figure 13-11. The X axis of the report shows the time of the temperature and the Y axis shows the temperature. Notice how it fluctuates as I mimicked a broken sensor by blowing different temperatures of air on the sensor via a hair dryer and air canister.

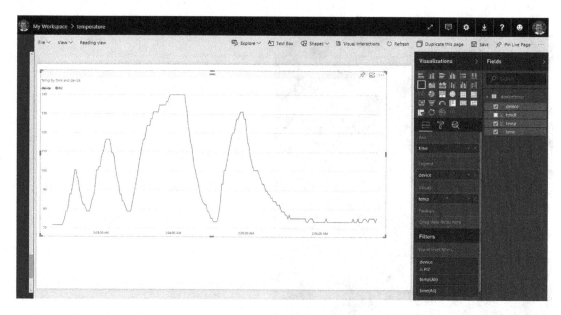

Figure 13-11. *Completed report*

Feel free to continue to experiment with the report by adding the humidity column or by adding additional sensors to the filter. At this point, you can save the report by clicking the Save icon in the upper right of the report. You will be prompted to provide a name for the report. Give it a good name and it will then be listed in your list of reports, as shown in Figure 13-12.

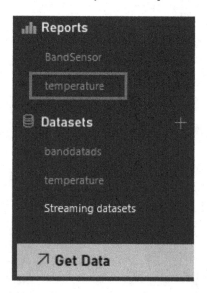

Figure 13-12. *Saved report*

OK, "temperature" isn't the best name for the report, but the point is that you now have a real-time report doing real-time analytics. Your homework assignment is to add additional line graphs to show temperatures from other devices. When you are done, you will have a nice-looking report.

The next part of your homework assignment is to add the report to a dashboard. While you are in the report, there is a Pin Live Page link at the top. Clicking that will open the Pin to Dashboard dialog, shown in Figure 13-13.

Figure 13-13. *Pinning the report to a dashboard*

The Pin to Dashboard dialog is where you specify which dashboard to pin the report to. What is awesome is that you can create a new dashboard through this if you haven't already created one. Very cool. So, what will you do? What will your dashboard look like? I am very interested in seeing your results so if you have created a nice-looking dashboard, feel free to email me a screenshot of your dashboard and I'll include it in my blog, highlighting your work. What's even more awesome and amazing is that with this dashboard you can share it and build a workspace so that others can see it!

Summary

This chapter focused on the real-time insights of your data. First, I discussed the value of hot path data and what real-time streaming and analytics provides. Then you learned how to gain those real-time insights via Azure Stream Analytics and Power BI by creating a Power BI account, and then modifying the Azure Stream Analytics job to add a new output to Power BI and changing the query to include the new output.

You then walked through building a Power BI report using the real-time data, starting with understanding the visualizations and using them to create a line chart report, and then using the fields from the dataset to filter and display the appropriate values and data in the report.

Lastly, you briefly walked through how to create a dashboard and add the report to the dashboard.

CHAPTER 14

■ ■ ■

Azure Machine Learning

Up to this point, the temperature data from your Raspberry Pis and Tessels have had quite the journey in Azure, including Azure Data Lake Store and Analytics, plus HDInsight—not to mention being streamed to Power BI in the last chapter. The data was processed using U-SQL and HDInsight, and routed using Azure Stream Analytics and Azure Data Factory, all with the purpose of doing processing and analytics on it to gain valuable insights.

But the journey for the data isn't quite over. Actually, some might say that the journey is just getting interesting. One of the goals of this book is to help the readers understand how the Azure data services can bring clarity and insight into existing and future data through analytical processing. But a big part of analyzing data is taking the next step and figuring out how to *learn* from the data.

The last chapter showed real-time reporting and analysis on temperature data. The chapter built a report and dashboard in Power BI in which sensor temperature data was displayed in real time. If an anomaly was detected, an alert could be triggered. Awesome. But how do we *learn* from that data?

Learning from data denotes that we use existing data to forecast future trends and outcomes, even behaviors. You and I see this everywhere. When you order a book from Amazon.com, the next time you visit the website you are presented with a list of recommended purchases that you might like based your recent purchases and other people who purchased the item you recently bought.

When my debit or credit card is used, my bank compares that transaction to other transactions to determine if any fraud is going on. When the data from my father's pacemaker is sent in for processing, that data is compared to data from other pacemakers to determine if the company needs to proactively make any changes or adjustments.

These systems *learn*; they use existing data to forecast and predict current and future outcomes. This learning is called *machine learning*, and it is used by data scientists to help applications and processes learn from existing data. The transaction at my bank uses machine learning to determine, based on learning from other transactions, if the current transaction is fraudulent.

This chapter is in no way a comprehensive look at machine learning nor am I a data scientist. Thus, this chapter is not intended to make you a machine learning expert. The goal of this chapter is to provide a glimpse at Azure Machine Learning and discuss where it fits into the ecosystem of Azure data services, and to show how to use it to look at real-time trends and outcomes, also called predictive analytics, from your data.

What Is Azure Machine Learning?

Machine learning is the process in which systems learn from existing data so as to predict and forecast future outcomes, and determine trends and behaviors. At the beginning of this chapter I mentioned both a financial and retail example. These are only two of myriad examples where machine learning is used. The better and faster a business can understand trends and behaviors, the faster they can improve their customer experience and thus their bottom line.

© Scott Klein 2017
S. Klein, *IoT Solutions in Microsoft's Azure IoT Suite*, DOI 10.1007/978-1-4842-2143-3_14

Azure Machine Learning brings the same predictive analytics capability to the cloud as a service, enabling data scientists to quickly create and deploy predictive models. Azure Machine Learning includes a large, built-in library of algorithms that can be used to design your model as well as a drag-and-drop studio for building and deploying models.

The following sections will walk through the steps necessary create and develop a predictive model, deploy the model as a web service, and then call that web service for real-time predictive analytics.

Creating the Azure Machine Learning Workspace

Working with Azure Machine Learning (ML) first requires that a workspace be created. These workspaces will contain all of the experiments you create within your Azure subscription. To create the workspace, select New ➤ Intelligence + Analytics, and then select Machine Learning Workspace, as shown in Figure 14-1.

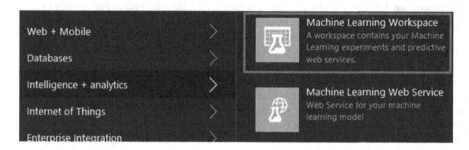

Figure 14-1. *Creating the ML workspace*

The Machine Learning Workspace blade will appear, which at the time of this writing looks like Figure 14-2. As usual, there are a few things to fill out, including a workspace name, resource group, location, web service plan, and pricing tier.

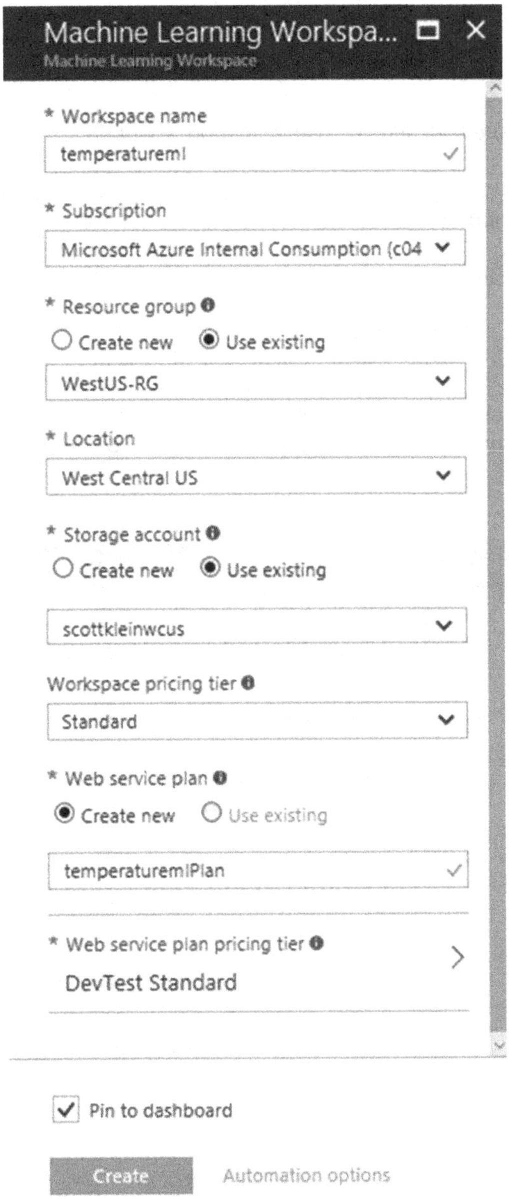

Figure 14-2. *Initial ML workspace configuration*

I will discuss the web service information later in the chapter, but suffice it to say now that these two pieces of information are necessary for when the predictive model is deployed as a web service. As such, this information is asking for the appropriate information when the model is deployed. Pricing for Azure Machine Learning is a bit different than other Azure services. Currently there are two pricing tiers (which apply to the use of Machine Learning Studio): free and standard. The free tier does not actually require an Azure subscription, which means you can use the Azure ML Studio without an Azure subscription. You also are limited to 100 modules per experiment and execution on a single node.

The standard tier is, in US dollars, $9.99/month. In this tier, an Azure subscription is required but most of the limitations of the free tier are removed, so you get unlimited modules in an experiment, unlimited stored (the free tier only provides 10GB), multiple execution nodes, and an SLA. Let me back up and say that the $9.99 is per seat per month.

There are a few other differences, so instead of listing them all here, you can read more about ML pricing here:

```
https://azure.microsoft.com/en-us/pricing/details/machine-learning/
```

Once all the information is filled out in the Machine Learning Workspace pane, click Create.

The Machine Learning Studio

The creation of the workspace takes less than a minute. Once created, you will be presented with the information about the workspace, as well as links for working with and managing Azure Machine Learning, as shown in Figure 14-3.

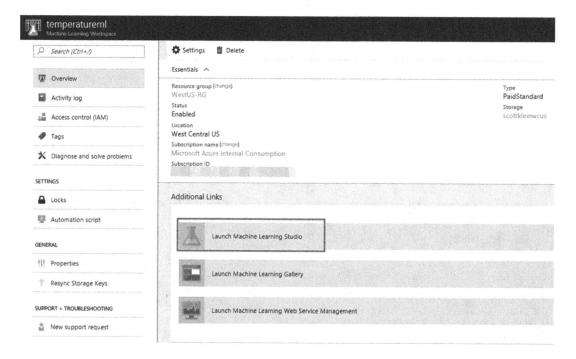

Figure 14-3. *Launching the Machine Learning Studio*

To start working with Azure Machine Learning and building a model, click the Launch Machine Learning Studio link, highlighted in Figure 14-3. The Azure Machine Learning Studio, shown in Figure 14-4, is where you build, test, and deploy your predictive analytics solutions. This is also where you upload your datasets and look at sample experiment models and datasets. In fact, before you create your own experiments and models, I highly recommend that you spend some quality time looking at some of the sample datasets and experiments. There are over 60 sample experiments and models that show predictive models on sentiment analysis, anomaly detection, fraud detection, demand estimation, and much, much more. In case you are wondering, datasets are stored in Azure Storage, and securely, which is why you need to choose an Azure Storage account during the Azure ML Workspace creation.

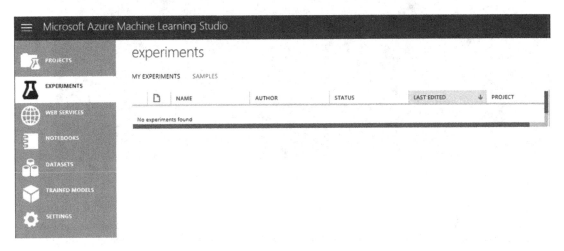

Figure 14-4. *The Azure Machine Learning Studio*

The Azure Machine Learning Studio is also where models are published as web services, which you will see later on in this chapter. These web services can then be consumed by custom applications or BI tools including Excel. As you can see in Figure 14-4, there is quite a bit that can be done in the studio, but for the purposes of this chapter, the focus will be on datasets, experiments, and web services. As a heads-up, the look and feel of Azure ML Studio was based on the previous version of the portal, so at some point the Studio will probably be updated and overhauled.

Before you can start building your experiment, you need to upload the data. If you have been following along, earlier chapters sent the data to Azure Blob Storage. This is exactly what you want, so go find one of those files in your Azure Blob Storage and download it. You can download the file directly from the Azure portal, or there are a number of tools that provide download capabilities from Azure Blog Storage, including Visual Studio.

You will need to rename the file to something meaningful. Once you have done that, you need to upload it as a dataset to the studio.

Uploading Your Data

To upload the file, click the + button in the lower left corner in the Machine Learning Studio. This will open up the New blade, shown in Figure 14-5. Select Dataset and then click the From Local File option, which will open up the dialog shown in Figure 14-6.

Figure 14-5. *Uploading a new dataset from a local file*

In the Upload a new dataset dialog, browse to the file you downloaded from Azure Blog Storage and renamed. I renamed my file SensorData.csv. The file you downloaded from Azure Blog Storage should already be a .csv, which is exactly what you want because that will allow you to select the "Generic CSV file with a header" option, as shown in Figure 14-6.

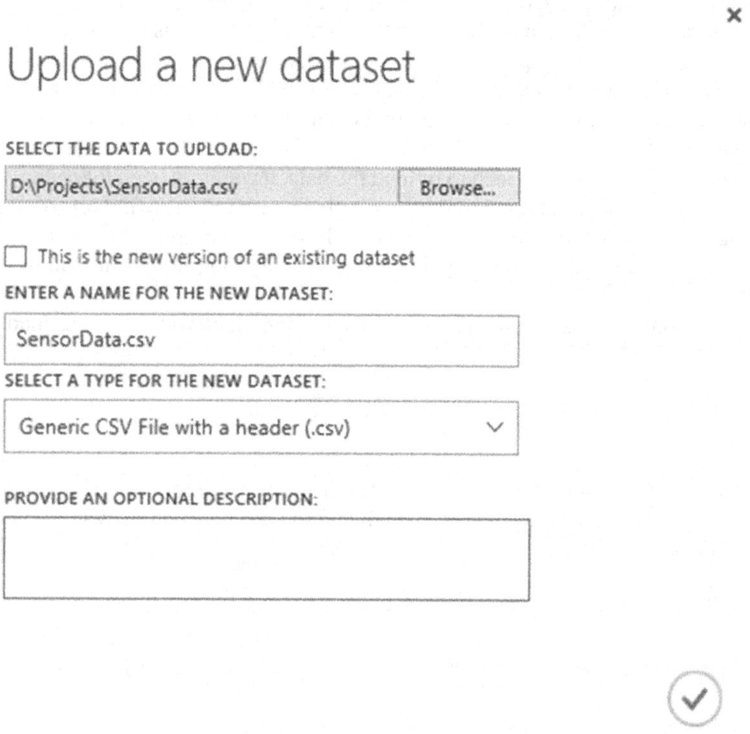

Upload a new dataset

SELECT THE DATA TO UPLOAD:

D:\Projects\SensorData.csv	Browse...

☐ This is the new version of an existing dataset

ENTER A NAME FOR THE NEW DATASET:

SensorData.csv

SELECT A TYPE FOR THE NEW DATASET:

Generic CSV File with a header (.csv)   ∨

PROVIDE AN OPTIONAL DESCRIPTION:

Figure 14-6. *Selecting the appropriate file and settings*

Click the checkmark in the lower right corner of the dialog to upload the dataset to the studio. Once uploaded, your dataset will be listed under the My Datasets tab, as shown in Figure 14-7.

Figure 14-7. *Temperature dataset*

Before moving on, I recommend again that you take some time looking at some of the sample datasets. If you look at some of the sample experiments, make sure to also look at the sample datasets that those sample experiments make use of.

Processing Temperature Data

With the dataset uploaded, the next step is to build the experiment. I will remind you here that the purpose of this chapter is not to make you an expert in data science or machine learning. There are plenty of books, tutorials, and other information out there that provide great information on these topics.

This section will build a very simplistic experiment that will be good enough to publish and walk through the remaining parts of the exercise, which is, again, the goal of this chapter. After creating the experiment, the following section will show how to publish it as a web service. The last section will then show how to call that web service from an Azure Data Factory pipeline to get real-time predictive analysis.

Creating the Experiment

At the beginning of this chapter you created the Machine Learning workspace, and the last section walked you through uploading the dataset. This section will walk through creating the experiment.

In the Machine Learning Studio, click New ➤ Experiment and select Blank Experiment, as shown in Figure 14-8.

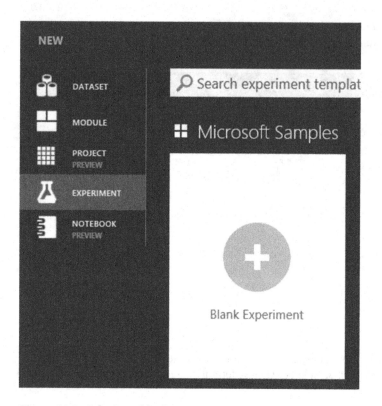

Figure 14-8. *Selecting a blank experiment*

Azure ML Studio will create a new blank experiment titled "Experiment created on" followed by today's date. You can change the name of the experiment by highlighting the entire default name (including the date) and giving the experiment a new name. For this example, name the experiment *iot_anomaly*.

The experiment can't be saved until a module is placed on the canvas so you need to do that. To the left of the canvas is a list of all the different modules organized by function, such as data transformations, data input and output, and so on. At the top is a node called Saved Datasets. Expand this node and you'll find two subnodes called My Datasets and Samples. I'll give you one guess under which node your recently uploaded dataset is located. Yes, expand the My Datasets node and there it is: your SensorData.csv dataset. Drag the dataset onto the canvas, at which point your canvas will look like Figure 14-9.

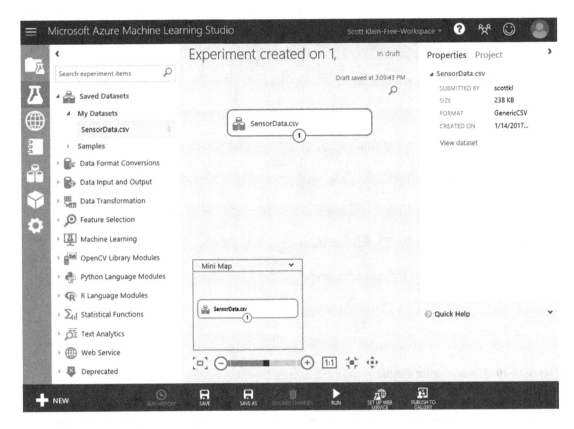

Figure 14-9. *Building the experiment*

You still won't be able to save the experiment because no actual modules have been added to the experiment (the dataset is not considered a module). Along the left of the list of modules are quick links to other facets of Machine Learning, including web services, datasets, existing trained models, and more. Along the right of the canvas is the Properties window, which is used to configure the properties of selected modules in the experiment. Along the bottom of the studio are buttons in which to save the experiment, run the experiment, and set up and publish the experiment as a web service. I point them out because you will be using the buttons along the bottom shortly.

Before adding a module, click the SensorData module on the canvas, and you'll see the lower point of the module turn to a number 1, as shown in Figure 14-9. Right-click the 1 and select Visualize from the context menu, which will display a pop-up with the dataset data along with additional information, as shown in Figure 14-10.

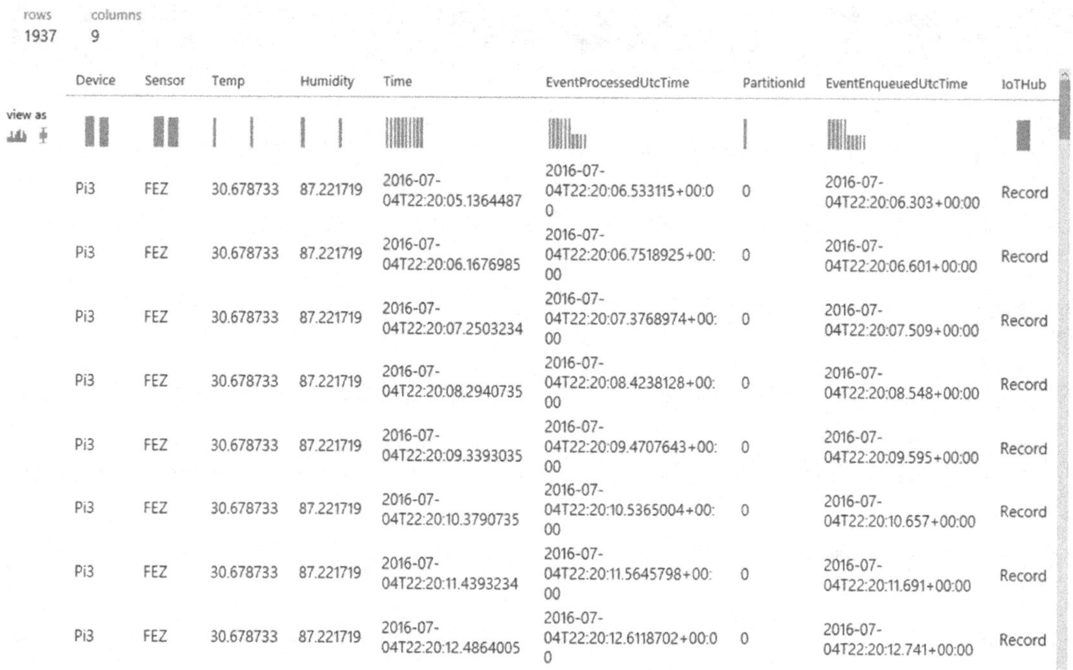

rows   columns
1937   9

	Device	Sensor	Temp	Humidity	Time	EventProcessedUtcTime	PartitionId	EventEnqueuedUtcTime	IoTHub
view as	Pi3	FEZ	30.678733	87.221719	2016-07-04T22:20:05.1364487	2016-07-04T22:20:06.533115+00:00	0	2016-07-04T22:20:06.303+00:00	Record
	Pi3	FEZ	30.678733	87.221719	2016-07-04T22:20:06.1676985	2016-07-04T22:20:06.7518925+00:00	0	2016-07-04T22:20:06.601+00:00	Record
	Pi3	FEZ	30.678733	87.221719	2016-07-04T22:20:07.2503234	2016-07-04T22:20:07.3768974+00:00	0	2016-07-04T22:20:07.509+00:00	Record
	Pi3	FEZ	30.678733	87.221719	2016-07-04T22:20:08.2940735	2016-07-04T22:20:08.4238128+00:00	0	2016-07-04T22:20:08.548+00:00	Record
	Pi3	FEZ	30.678733	87.221719	2016-07-04T22:20:09.3393035	2016-07-04T22:20:09.4707643+00:00	0	2016-07-04T22:20:09.595+00:00	Record
	Pi3	FEZ	30.678733	87.221719	2016-07-04T22:20:10.3790735	2016-07-04T22:20:10.5365004+00:00	0	2016-07-04T22:20:10.657+00:00	Record
	Pi3	FEZ	30.678733	87.221719	2016-07-04T22:20:11.4393234	2016-07-04T22:20:11.5645798+00:00	0	2016-07-04T22:20:11.691+00:00	Record
	Pi3	FEZ	30.678733	87.221719	2016-07-04T22:20:12.4864005	2016-07-04T22:20:12.6118702+00:00	0	2016-07-04T22:20:12.741+00:00	Record

Figure 14-10. Visualizing the dataset

If you are familiar with R, then this will look familiar to you. The pop-up shows the dataset data as well as key dataset metrics displayed along with histograms of the values and statistical details of the ranges including average values as well as variance standard deviation.

Close the pop-up. Let's add some modules to the experiment. From the list of modules, expand the Data Transformation node, and then expand the Manipulation subnode and drag the Select Columns in Datset and Edit Metadata modules onto the canvas below the SensorData.csv.

Connect the dataset to the Select Columns in Dataset module by clicking the output port of the dataset and connecting it to the input port of the Select Columns in Dataset module. Connect the Select Columns in Dataset module to the Edit Metadata module the same way.

The Select Columns in Dataset module should have a red exclamation in it, so click that module, at which point the Properties has a "Launch column selector" button. Click the "Launch column selector" button, which opens the Select columns dialog. You are only really interested in three columns, Device, Temp, and Humidity, so move them to the right and click the checkmark, as shown in Figure 14-11.

Select columns

×

BY NAME	AVAILABLE COLUMNS		SELECTED COLUMNS
WITH RULES			

AVAILABLE COLUMNS

All Types ∨   search columns   🔍

Sensor
Time
EventProcessedUtcTime
PartitionId
EventEnqueuedUtcTime
IoTHub

>

<

6 columns available

SELECTED COLUMNS

All Types ∨   search columns   🔍

Device
Temp
Humidity

3 columns selected

✓

Figure 14-11. Selecting columns to include

The Device column has a set of discrete values whereas the other two columns represent continuous values that can occur on a scale range. Thus, you need to select the Device column and make it categorical. Click the Edit Metadata module and again click the "Launch column selector" button. In the Select columns dialog, select the Device column from the list of columns and click OK. Back in the Properties window, set the Categorical option to Make categorical, as shown in Figure 14-12.

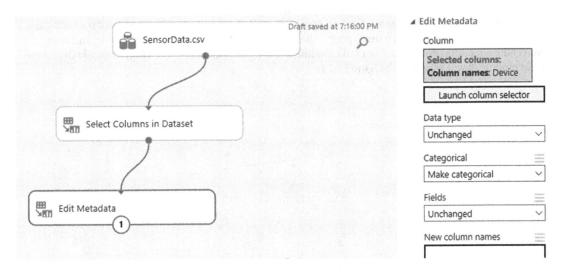

Figure 14-12. Selecting the desired columns of the dataset

Your experiment should now look like Figure 14-13.

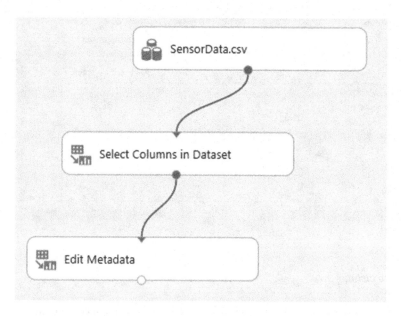

Figure 14-13. *First part of the experiment*

Next, from the Data Transformation ➤ Sample and Split nodes, drag the Split Data module on the canvas. Then, from the Machine Learning ➤ Initialize Model ➤ Anomaly Detection node, drag the One-Class Support Vector Machine module onto the canvas. Next, from the Machine Learning ➤ Train node, drag the Train Anomaly Detection Model module onto the canvas.

Connect the Edit Metadata module to the Split Data module, connect the output port of the One-Class Support Vector Machine to the left input port of the Train Anomaly Detection Model module, and then connect the left output port of the Split Data module to the right input port of the Train Anomaly Detection Model module. It will appear as in Figure 14-14.

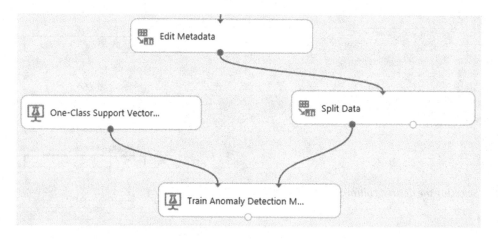

Figure 14-14. *Adding to the experiment*

In the Properties page for the Split Data module, update the fraction of rows in the first output dataset from 0.5 to 0.7. This allows you to split between training and testing data to 70%. The data split at 70%, or 0.7, specifies that 70% of the data will be used to "train" the model to detect any anomalies in the device data, meaning that if humidity or temperature are out of expected bounds for a particular device you should be able to detect this.

When you have trained your model, you will know what to look for when new data comes in. With this in mind, you'll use the other 30% of the data in your initial dataset to "test" the effectiveness of your predictions.

To do this, drag the Score Model module onto the canvas from the Machine Learning ➤ Score nodes. Connect the output port of the Train Anomaly Detection Model module to the left input port of the Score Model module, and then connect the right output port of the Split Data module to the right input port of the Score Model module, as shown in Figure 14-15.

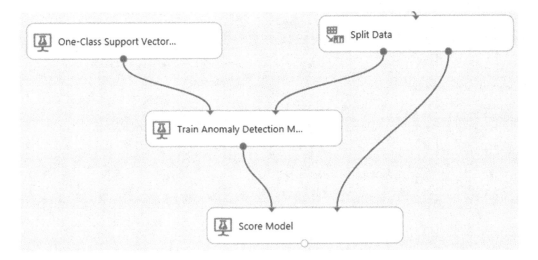

Figure 14-15. *Completed experiment*

If you haven't saved the experiment yet, click the Save button to save all your hard work. If you renamed your experiment, it will be now saved with that name. Hopefully you named it something you remember (I named mine iot_anomaly) because you will be looking for it later.

Once you have saved the experiment, press the Run button to run the experiment. Running the experiment does an initial training and scoring of the data and will take only a minute. At this point, your experiment should look like Figure 14-16. Once the run is complete, all the modules will have a green check mark to show that that module ran successfully (as shown in Figure 14-16).

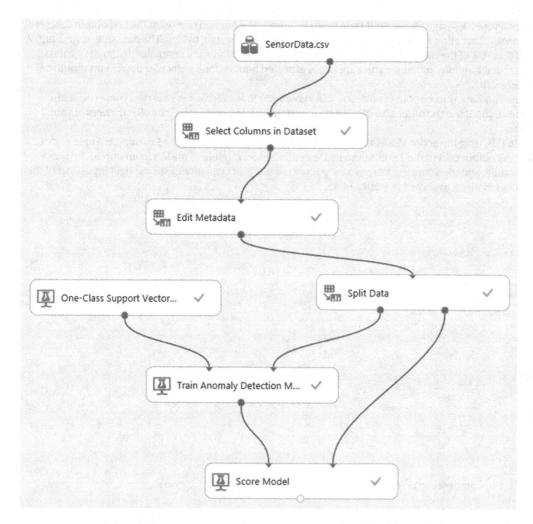

Figure 14-16. *Trained model*

As well, now that the run is complete, you can see the scored data by looking at the output of the Score Module. Just like you did on the dataset, click the Score Model module and the output will turn to a number 1. Right-click that and select Visualize. As shown in Figure 14-17, the scored output adds two more columns to your dataset: Scored Labels and Score Probabilities. The Score Label is a 1 or a 0, where a 1 represents an anomalous reading.

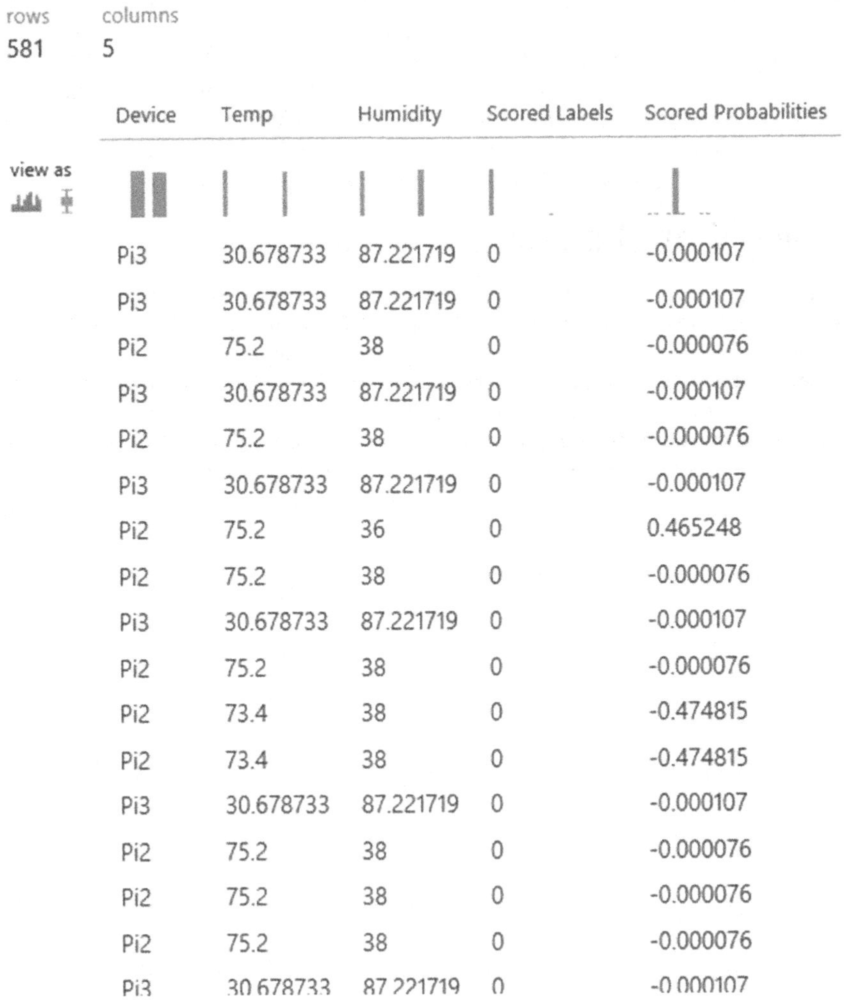

rows   columns
581   5

	Device	Temp	Humidity	Scored Labels	Scored Probabilities
view as					
Pi3	30.678733	87.221719	0	-0.000107
Pi3	30.678733	87.221719	0	-0.000107
Pi2	75.2	38	0	-0.000076
Pi3	30.678733	87.221719	0	-0.000107
Pi2	75.2	38	0	-0.000076
Pi3	30.678733	87.221719	0	-0.000107
Pi2	75.2	36	0	0.465248
Pi2	75.2	38	0	-0.000076
Pi3	30.678733	87.221719	0	-0.000107
Pi2	75.2	38	0	-0.000076
Pi2	73.4	38	0	-0.474815
Pi2	73.4	38	0	-0.474815
Pi3	30.678733	87.221719	0	-0.000107
Pi2	75.2	38	0	-0.000076
Pi2	75.2	38	0	-0.000076
Pi2	75.2	38	0	-0.000076
Pi3	30.678733	87.221719	0	-0.000107

Figure 14-17. *Scored data*

Your model isn't trained yet, so the next step is to "train" it. Right-click the Train Anomaly Detection Model module and select Trained Model ➤ Save as Trained Model. In the Save Trained Model dialog, make sure the property "This is a new version of an existing trained model" is unchecked and then provide a name for the new trained model. There will be a default name provided, but you can erase it and provide a new name. For this example, name the trained model the same thing as your experiment, meaning iot_anomaly if you named it the same as I did.

Click Save, at which point the new trained model will be listed in the Trained Models node at the left. This is one of several ways a trained reusable model can be created and is now an artifact for you. As such, you can now delete the One Class Support Vector Machine and Train Anomaly Detection Model modules from the experiment, and drag the new iot_anomaly trained model onto the canvas it their place. Connect the output of the iot_anomaly module to the left input of the Score Model module. Rerun the model to make sure all is well.

As a quick note, there are two types of anomaly detections: the One-Class Support Vector Machine, which was used in the example, and the Principal Component Analysis, or PCA-based Anomaly Detection. Both of these anomaly detection modules can be used when it is easy to get training from one class but difficult to get good samples of the targeted anomalies. As such, you can test both modules side by side to see which one works better for you, which Azure ML is great at! Your model is now complete and ready to be published as a web service so that it can be called form Azure Data Factory as data comes in.

Machine Learning Web Service

You created a Machine Learning model so that you can easily detect sensor temperature anomalies as data comes in real time. The last section created a Machine Learning model to capture those anomalies. The next step is to publish that model as a web service that will enable you to call it as data comes in.

With your model open in ML Studio, click the Set up Web Service button, which is next to the Run button. This process adds two new modules automatically to the canvas: the Web Service Input and Web Service Output modules. As you can see in Figure 14-18, your input isn't the SensorData.csv dataset any more, but the input from your web service input endpoint, which is connected directly to the Select Columns in Dataset module.

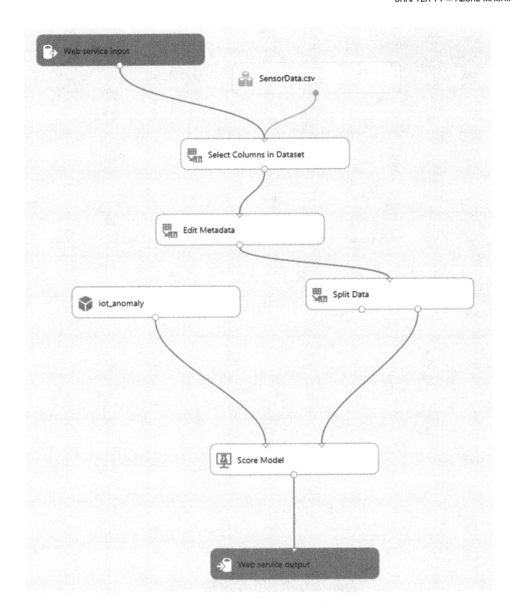

Figure 14-18. *Adding web service input and output*

Also, the Score module outputs data to the Web Service Output module so that you can retrieve the output data. The Web Service Input and Web Service Output modules simply allow you to send input to your model from a web service endpoint and feedback output to the calling code.

Because of the work you did earlier in training your model, you can move the links around a little since you only want to send the three column values to the web service, not all the values. To do that, delete the link from the Web Service Input to the Select Columns in Dataset and then reconnect the web service input directly to the right input of the Score module, as shown in Figure 14-19.

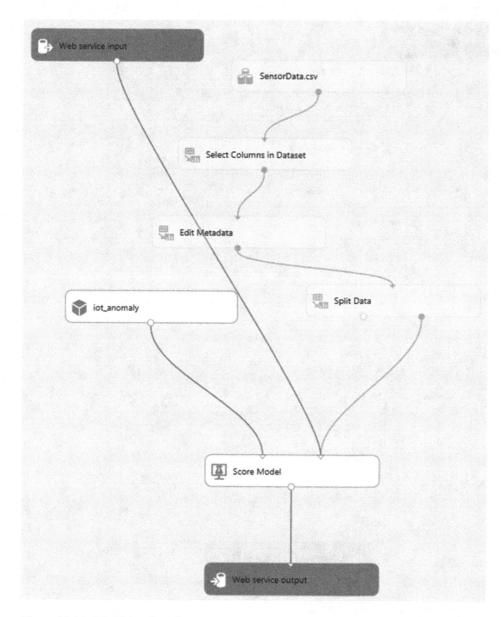

Figure 14-19. *Modifying the web service input*

Now that everything is linked up again, click Run again and run the experiment. When all the green ticks in all the various boxes are done, it is time to deploy the web service. Notice that the Set up Web Service button is now named Deploy Web Service. Click the Deploy Web Service button, which will open a dashboard to allow you to create the web service and also tell you everything about the web service calls.

In the dashboard, shown in Figure 14-20, copy the API key to Notepad or somewhere you can get to it shortly. The API key is used to authenticate to the web service and even though you will have an endpoint URL, you will also need to pass the key to authenticate.

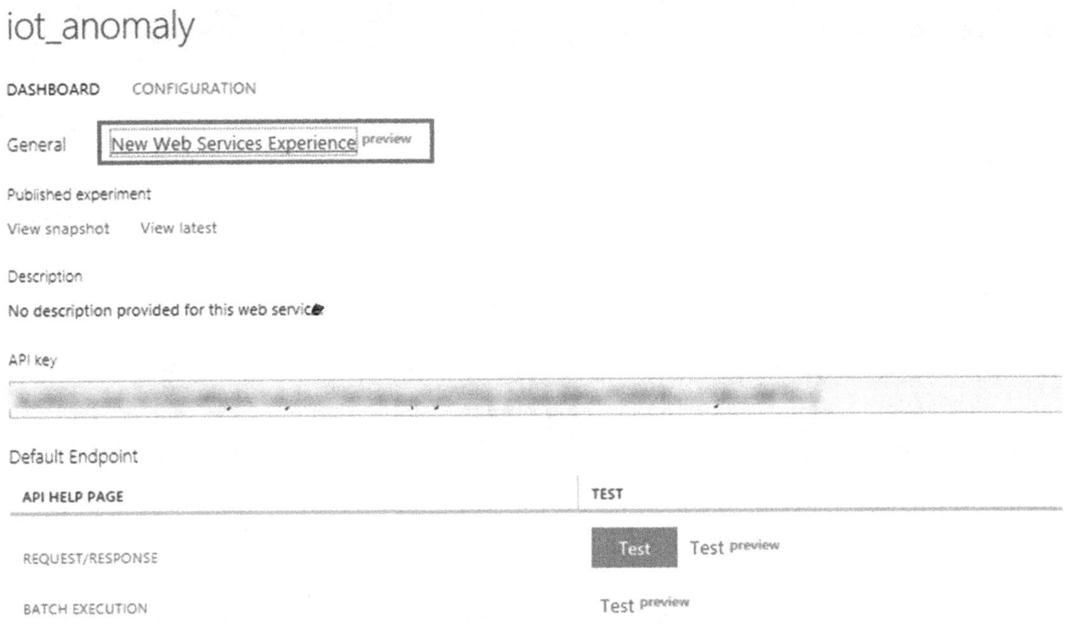

iot_anomaly

DASHBOARD   CONFIGURATION

General   New Web Services Experience preview

Published experiment

View snapshot   View latest

Description

No description provided for this web service

API key

Default Endpoint

API HELP PAGE	TEST
REQUEST/RESPONSE	Test Test preview
BATCH EXECUTION	Test preview

Figure 14-20. *The Deploy Web Service dashboard*

Next, click the New Web Services Experience link on the dashboard, which will open a new browser page, shown in Figure 14-21. This page is a quick start page that, among other things, provides the ability to create the necessary endpoints, and test and use those endpoints.

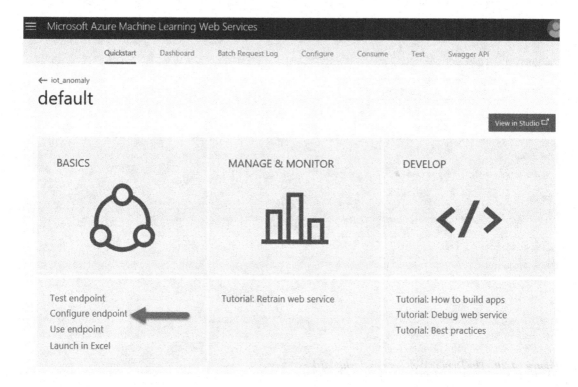

Figure 14-21. *Machine Learning web services quick start page*

In the Basics section, click the Configure endpoint link and configure the endpoint information. First, you need to provide a description. Second, select the logging level. As an FYI, when you created the Machine Learning workspace, you should have had the choice of selecting the free or standard pricing tier. With the free tier, only the "None" logging option is available. Logging is only available in the standard pricing tier. Next, make sure the Sample Data Enabled option is set to No.

Once your endpoint information is configured as shown in Figure 14-22, click Save.

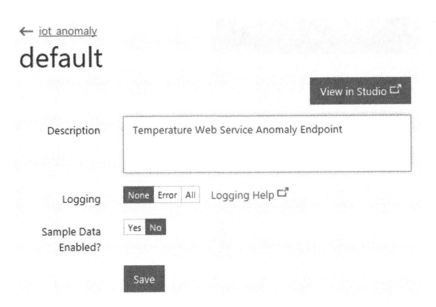

Figure 14-22. *Web service endpoint configuration*

After you have saved the endpoint configuration, go back to the quick start page and click the "Use endpoint" link shown in Figure 14-23, which will open the Web Service Consumption Options page.

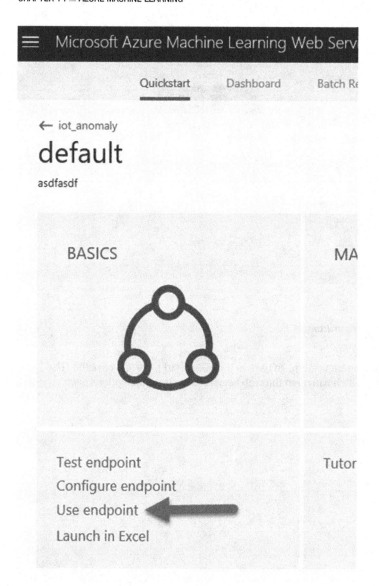

Figure 14-23. *Clicking the "Use endpoint" link*

In the Web Service Consumption Options page, shown in Figure 14-24, there are a few items of information available but the piece of information you want is the batch requests endpoint URL link, highlighted in Figure 14-24. However, I highly recommend you spend some time looking at the tutorials and other useful information available on this page. Copy the batch requests URL to Notepad or somewhere you can easily get to it because you will need it shortly.

Figure 14-24. *Retrieving the batch requests endpoint URL*

At this point, the web service is published and ready to be used. You can verify that it has indeed been published by going back to the ML Studio and clicking the Web Services icon on the left navigation bar.

OK, you are almost done. It is time to put the machine learning model and web service to use. The next section will show you how to call the web service using Azure Data Factory.

Azure Data Factory

Chapters 7 and 8 introduced you to Azure Data Factory and walked you through creating a pipeline and orchestrations in order to automate data movement and data transformation. This section will go back and visit Azure Data Factory to be able to call the web service published in the last section in order to pass data into the Machine Learning model.

In the Azure portal, open the Data Factory created in Chapter 8 called TemperatureDF (unless you called it something different). In the Overview pane, click the Author and Deploy tile to open the familiar authoring blades.

The first step is to create a new linked service for the web service created earlier and to do that, if you remember, is to click the More link, select New Compute, and then select the Azure ML option, as shown in Figure 14-25.

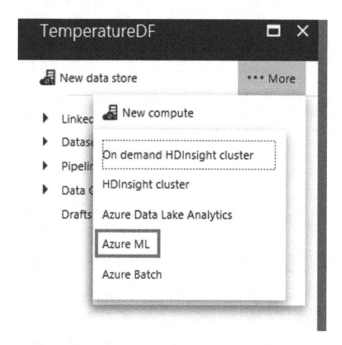

Figure 14-25. *Creating a New Azure ML linked service*

Replace the default code for the Azure ML linked service with the following code. However, here is where the information you copied earlier to Notepad (or wherever you saved it) comes in. Copy the API Key into the apiKey value, and replace the web service endpoint URL below with the one you copied. Deploy the linked service once you have updated all the information.

```
{
    "name": "AzureMLTempAnomalyEndpoint",
    "properties": {
        "description": "",
        "hubName": "temperaturedf_hub",
        "type": "AzureML",
        "typeProperties": {
            "mlEndpoint": "https://ussouthcentral.services.azureml.net/workspaces/944629a4654748
                7dbf2a6f71ef3bd3f2/services/9a5e3a4eef744300b1c34e36918a94ee/jobs?api-version=2.0",
            "apiKey": "**********"
        }
    }
}
```

The final step in all of this is to create the pipeline. Again, click the More link and select New Pipeline. Replace the default code with the following code. This code defines a pipeline with a single activity, an Azure ML batch scoring activity with an input and an output:

```
{
    "name": "AzureMLBatchScore",
    "properties": {
        "description": "Description",
```

```
    "activities": [
        {
            "type": "AzureMLBatchScoring",
            "typeProperties": {},
            "inputs": [
                {
                    "name": "AzureBlobDataset"
                }
            ],
            "outputs": [
                {
                    "name": "AzureSQLDataset"
                }
            ],
            "policy": {
                "timeout": "01:00:00",
                "concurrency": 1,
                "executionPriorityOrder": "NewestFirst",
                "retry": 3
            },
            "scheduler": {
                "frequency": "Month",
                "interval": 1,
                "style": "StartOfInterval"
            },
            "name": "AzureMLScoringActivityTemplate",
            "linkedServiceName": "AzureMLTempAnomalyEndpoint"
        }
    ],
    "isPaused": false,
    "hubName": "temperaturedf_hub",
    "pipelineMode": "Scheduled"
    }
}
```

The input is the Azure Blob dataset and the output is the Azure SQL dataset, meaning the pipeline will pick up data from the Azure Blob Storage account and call the web service, passing the data to the web service input endpoint. If you recall, the data in Azure Data Storage came from Azure Stream Analytics. See the flow here?

Data is sent from the devices to Azure IoT Hub, picked up by Azure Steam Analytics, and routed to Azure Blob Storage, where it is picked up by Azure Data Factory and sent to the Azure ML web service for anomaly detection, the results of which are dropped into the Azure SQL Database.

So click Deploy to save and deploy the pipeline. Once deployed, you can go back to the Azure Data Factory Overview pane and click the Diagram tile, which will show you the new pipeline, similar to the one in Figure 14-26.

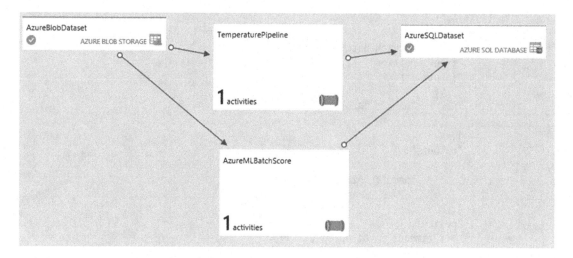

Figure 14-26. *Azure ML Batch Score pipeline*

Granted, the pipeline defined above runs once a month. Realistically you would want that data processed much more frequently to detect anomalies, so you could change the scheduler frequency to minute or hour, depending on need.

Whew! That was a lot of work, but hopefully you see the value in this chapter and of getting real-time predictive analytics of data. Again, this chapter was not intended to go deep in Azure Machine Learning, but simply to show you the possibilities via an example. I highly recommend that you continue to play with the machine learning model and with the pipeline. You might consider changing the input to pick the data up from Azure Data Lake Store.

Summary

The goal with this chapter was to provide insight into how to gain real-time predictive analytics on your data in order to look for trends and patterns. To do so, you walked through an example of using Azure Machine Learning and Azure Data Factory to process data in real time.

You began by creating an Azure Machine Learning workspace and uploading your sensor data as a dataset to it. Then you created a Machine Learning experiment, saved it as a trained model, and published the model as a web service.

Lastly, you created a new workflow pipeline in Azure Data Factory to call the web service for anomaly detecting.

PART IV

More on Cortana Intelligence

CHAPTER 15

■ ■ ■

Azure Data Catalog

Throughout this book, you have learned that data can come from many different sources and many different formats. Specifically speaking, the source of the data for this book has come from a number of devices and sensors. The different sources and formats are two of the three Vs mentioned at the beginning of this book: variety and volume.

You have also learned that this data can be stored in a number of different data stores in Microsoft Azure, including Azure Data Storage, Azure Data Lake Store, Azure SQL Database, and more. But these data stores are just a few of the data store options in the cloud and within your organization.

Looking at this through multiple lenses of a DBA and an end user, how do you know what data is in your organization or company and the value it has? How do you figure out and understand all the different sources of data and, more importantly, even remotely begin to gain insight into the data?

Assume for a moment that you are starting a new job today as a DBA or someone who needs access to critical data via reports or other means. Figuring out all the different data sources and what data those data sources contain is a challenge. Typically, users (whether a DBA or end user) have no idea a data source exists until they either discover it by accident or via another process.

The solution to all of these challenges is a central location where users can discover data sources and understand the data that is in those data sources easily so that they can get more value from existing information. This is where Azure Data Catalog comes in.

What Is Azure Data Catalog?

Data discovery can be looked at from two sides; those producing the data and those consuming the data. Data producers have the challenge of securing the data: admitting those who need to have access and restricting those who don't need access. They also have the challenge of documenting and annotating data. Today, if I want to know what data is in a folder, I look at the folder name, which is not the best documentation.

Data consumers also have challenges, and a few were mentioned earlier. Data discovery challenges for those looking for data are typically in the class of not knowing data exists or where to go looking for it. Once found, the consumer may or may not have access to the data and, if they do, they spend more time browsing the data to understand it and its intended use. The frustration doesn't stop there because if they have a question about the data, the new challenge becomes one of finding the owner of the data.

Azure Data Catalog was created to address every single one of these challenges and more. As a fully-managed cloud service (just like all of the other Azure services), the intent and goal of Azure Data Catalog is to make data discovery simple, help users easily understand the discovered data, and to get the best value possible out of said discovered data.

Azure Data Catalog doesn't move any data around. Instead, it keeps the data in its existing location but tags the data with metadata, which is in turn stored in Azure Data Catalog for easy discoverability. Azure Data Catalog also allows users to contribute to the catalog by tagging and annotating data sources that have

© Scott Klein 2017
S. Klein, *IoT Solutions in Microsoft's Azure IoT Suite*, DOI 10.1007/978-1-4842-2143-3_15

already been registered to further enrich the discovery capabilities, and to register new data sources for discovering. At its core, Azure Data Catalog is an API like the other services in Azure, which provide many benefits for working and integrating with Azure Data Catalog.

Scenarios

The most basic scenario is that of plain and simple data discovery: finding what data is out there, what data do you need to do your job, and who owns it. Beyond this scenario there is one other scenario that you probably would not think of, and this of self-service business intelligence.

Self-service BI lets users create their own reports and dashboards without waiting or relying on different teams or organizations to do the development. In BI scenarios, it is common when building reports and dashboard for data to be pulled in from multiple sources. The majority of these sources are known, but at times some of these data sources are not known.

Azure Data Catalog helps in the effort of not only the data discovery of the multiple data sources but to allow users to build their reports and dashboards from the discovered data sources easily. From there, the users can contribute to the growth of the catalog and add value to the existing data sources.

So, with this background, the rest of this chapter will build a data catalog, register a couple of datasets, and then connect to them.

Working with Azure Data Catalog

The following sections will take a detailed look at working with Azure Data Catalog. As you learned earlier, Azure Data Catalog makes it easy to discover different data sources. For the purposes of the example in this chapter, I created a couple of data sources for you to register. As shown in Figure 15-1, I created an Azure SQL Database and an Azure Data Lake Store account. The Data Lake Store account has a single file that contains the sensor data. The Azure SQL Database contains a single table that contains supporting data. Very simple. The AdventureWorks sample database can be downloaded from Codeplex (http://msftdbprodsamples.codeplex.com/).

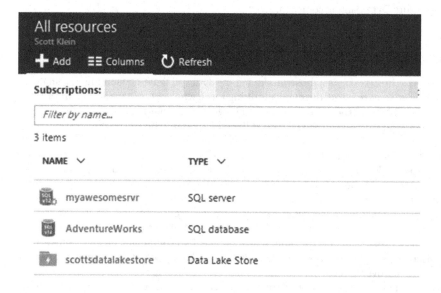

Figure 15-1. *Data sources*

Let's get started. First, you'll create the Data Catalog and register the data sources. The following sections will walk through that.

Provision Azure Data Catalog

This section will be pretty short, and I'll explain why. There are a couple of details that I will point out that will save you a lot of time. First, each Azure Data Catalog requires an organizational account. Not a Microsoft account such as a Live ID, but an account that is tied to your organization with Active Directory. Second, only a single Azure Data Catalog can be created per organization. Note that I said "organization," not Azure subscription.

The reason for this restriction is that Azure Data Catalog is intended to be a system of records for all data sources across the enterprise. OK, that makes sense, but at this point you are asking yourself "How can I play around with and test Azure Data Catalog at home?" Great question, and luckily there is an answer.

SQL Server MVP Melissa Coates wrote a blog post that walks through how to do this using a Microsoft account (for example, a Live ID). It essentially entails creating an Azure Active Directory Account, allowing that account to be co-administrator in your Azure subscription, and then signing into the Azure Data Catalog portal using the new Azure Active Directory account. So please read Melissa's awesome blog post:

www.sqlchick.com/entries/2016/4/20/how-to-create-a-demo-test-environment-for-azure-data-catalog

Follow the blog post step by step and you will have an Azure Data Catalog up and running in just a few minutes. Now, I did try this in the new portal since Azure Active Directory is supported in the new portal but I think there must have been a hiccup or something because it let me create an Azure Active Directory but then it didn't show up in my list. So, feel free to try it in the new portal. I haven't reached out to the PMs yet on this issue but I will do so to see if it is just me or if this is a bug. However, it works in the old portal so if you run into issues, hop over to the old portal and have at it.

When you are done, be sure to thank Melissa for the awesome blog post and then pop back over to this chapter to pick up where you left off, which is the next section on registering data sources in your new Data Catalog.

Registering Data Sources

So you have your Azure Data Catalog created. The next step is to register some data assets, which is the process of pointing Azure Data Catalog at a data source and extracting key metadata from the data source including names, location, and other vital information. Part of the registration process copies that metadata into the catalog but the data assets remain in their original location. Copying over the metadata into the Data Catalog is which makes the data sources more discoverable and easier to understand.

To register a data source, open your favorite browser and navigate to the Azure Data Catalog home page at https://azuredatacatalog.com. The home page of the Azure Data Catalog includes several links along the top to publish data and configure settings for the Data Catalog. The home page also shows any registered and pinned assets as well as any saved searches for quick access.

Since you have no registered assets, your first step is to defining a data source and register assets. To do this, you can either click the big blue Publish Data button in the middle of the home page or click the Publish button near the top right. Clicking either will take you to the Publish page, shown in Figure 15-2.

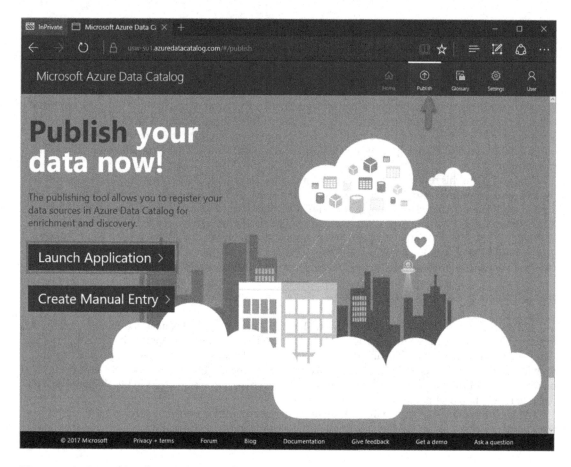

Figure 15-2. Launching the registration tool

On this page you can create a manual entry, which I won't walk through here, but the easier choice is to click the big blue Launch Application button. The application is a simple Windows application that runs on the local computer and allows you to register a data asset. So, go ahead and click the Launch Application button.

The registration tool will download and install. The download will take a minute or two and then install. You may need to click Accept as part of the install. Once installed, it will automatically run and you will be presented with the Welcome page.

On the Welcome page, click the Sign In button, which will pop up the Sign in dialog, shown in Figure 15-3. In this dialog, use the Active Directory user you created as part of the Azure Data Catalog creation. Since you are registering data assets with Azure Data Catalog, you need to authenticate via the registration tool with Azure Data Catalog.

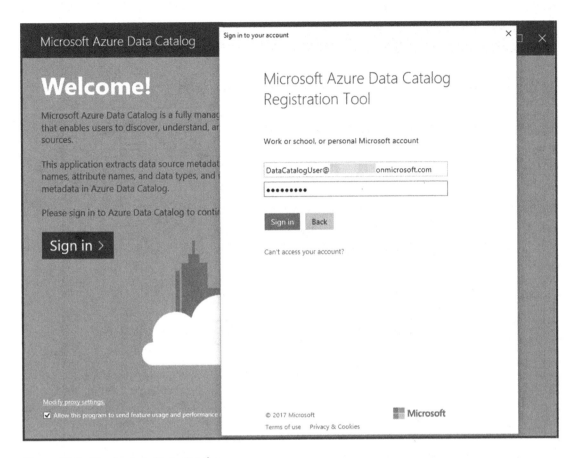

***Figure 15-3.** Signing in to Data Catalog*

Once authenticated, you will be presented with a list of data sources to connect to and from which to register data assets. As you can see from Figure 15-4, there are currently 17 options that range from relational to non-relational data sources, including SQL Server, SQL Data Warehouse, Hadoop, Azure DocumentDB, Azure Table and Blob Storage, and more. Users of Azure Data Catalog can work with other supported data sources through APIs, manual entry, or through the registration tool. I won't list them all here, but the following URL shows all the data sources supported today and through which method they can be accessed and published.

`https://docs.microsoft.com/en-us/azure/data-catalog/data-catalog-dsr`

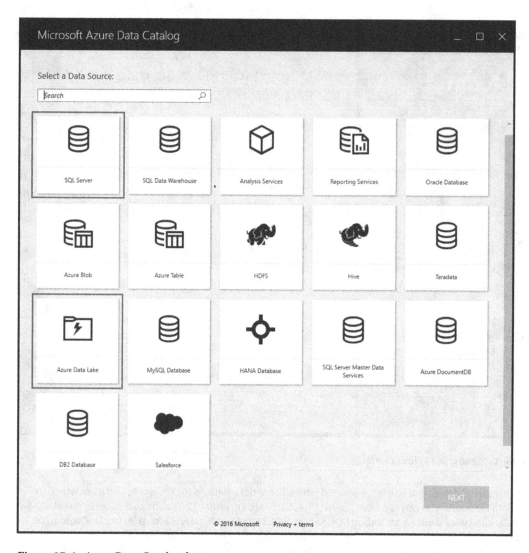

Figure 15-4. *Azure Data Catalog data sources*

For this example, select SQL Server (which also allows you to connect to Azure SQL Database) and Azure Data Lake Store. I will walk you through registering data assets from Azure SQL Database, and your homework assignment will be to do the same for Azure Data Lake.

In the Select a Data Source page, shown in Figure 15-4, click SQL Server and then click Next. In the Server Connection page, shown in Figure 15-5, enter the server information and click Connect. For this example, I created an Azure SQL Database and associated server. The server is called myawesomeserver, and when I created the database, instead of creating a blank database, I elected to use the sample AdventureWorksLT database.

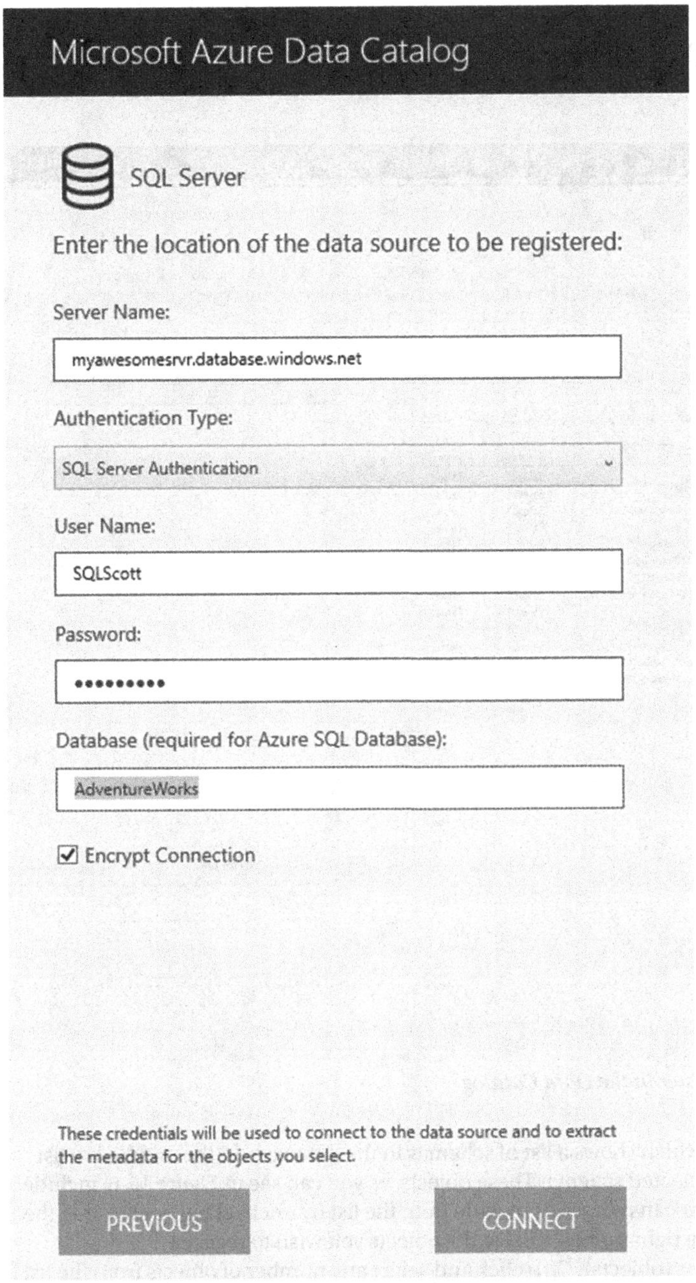

Figure 15-5. *Connecting to Azure SQL Database*

Also, since I am using Azure SQL Database in this example, I am using SQL Server Authentication instead of Windows Authentication (since my Azure SQL Database is not part of an Active Directory). You can read more about connecting to Azure SQL Database or SQL Data Warehouse via Azure Active Directory at https://docs.microsoft.com/en-us/azure/sql-database/sql-database-aad-authentication.

Once the information is filled out, click Connect.

After clicking Connect, the next page will show the list of objects from which you can register assets, as shown in Figure 15-6.

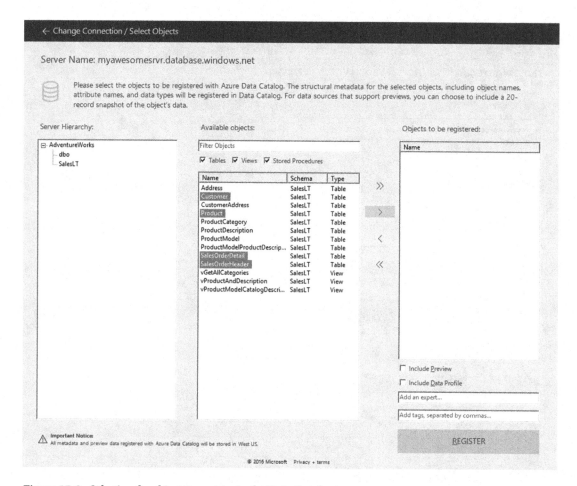

Figure 15-6. *Selecting the objects to register in the Data Catalog*

On the left is the server hierarchy, which shows a list of schemas in the database. In the middle is a list of available objects that pertain to the selected schema. These objects, as you can see in Figure 15-6, include tables, views, and stored procedures. You can remove or include from the list by unchecking or checking the appropriate checkbox above the list. The right section will list the objects you wish to register.

To register the metadata of the assets (objects), Ctrl+click and select any number of objects from the list of objects. In this example, I have selected the Customer, Product, SalesOrderDetail, and SalesOrderHeader tables. Once selected, click the selected arrow to move those objects to the "Objects to be registered" list, as shown in Figure 15-7.

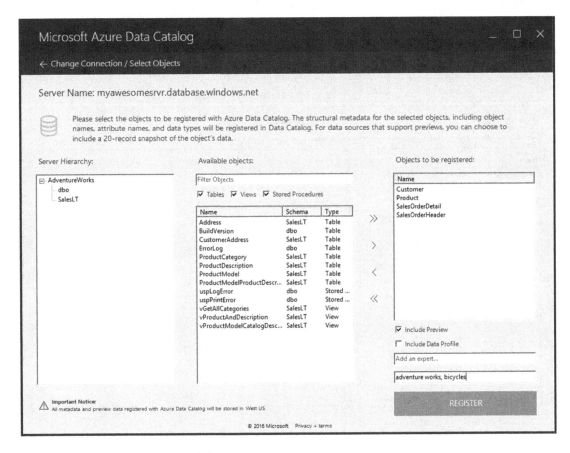

Figure 15-7. *Adding tags to selected objects*

Here is where the fun begins. Before clicking the Register button, now is the time to add tags. In the Add tags text box, enter a few tags that will help users find the data source. For example, I entered *bicycles* and *adventure works*. Adding tags adds search tags for the selected assets in the list. The tags should be descriptive and make it easy for users to find the data source and identify the type of data in the data source.

Also, be sure to check the Include Preview checkbox. This will include a snapshot preview of the data when registering the data source. Checking this checkbox will copy up to 20 records from each table selected and copy them to the data catalog, which will allow users to get a quick visual as to the type of data in the asset. The Include Data Profile option will include a snapshot of the statistics for each object. For this example, I left that option unchecked.

Once you are finished on this page, click Register. Azure Data Catalog will then go through the object selected and register that asset. Figure 15-8 shows the registration process. Depending on the number of assets selected, the registration process could take a while. Luckily, there is a percent completion on the page so you know roughly how long it will take and the status of the process.

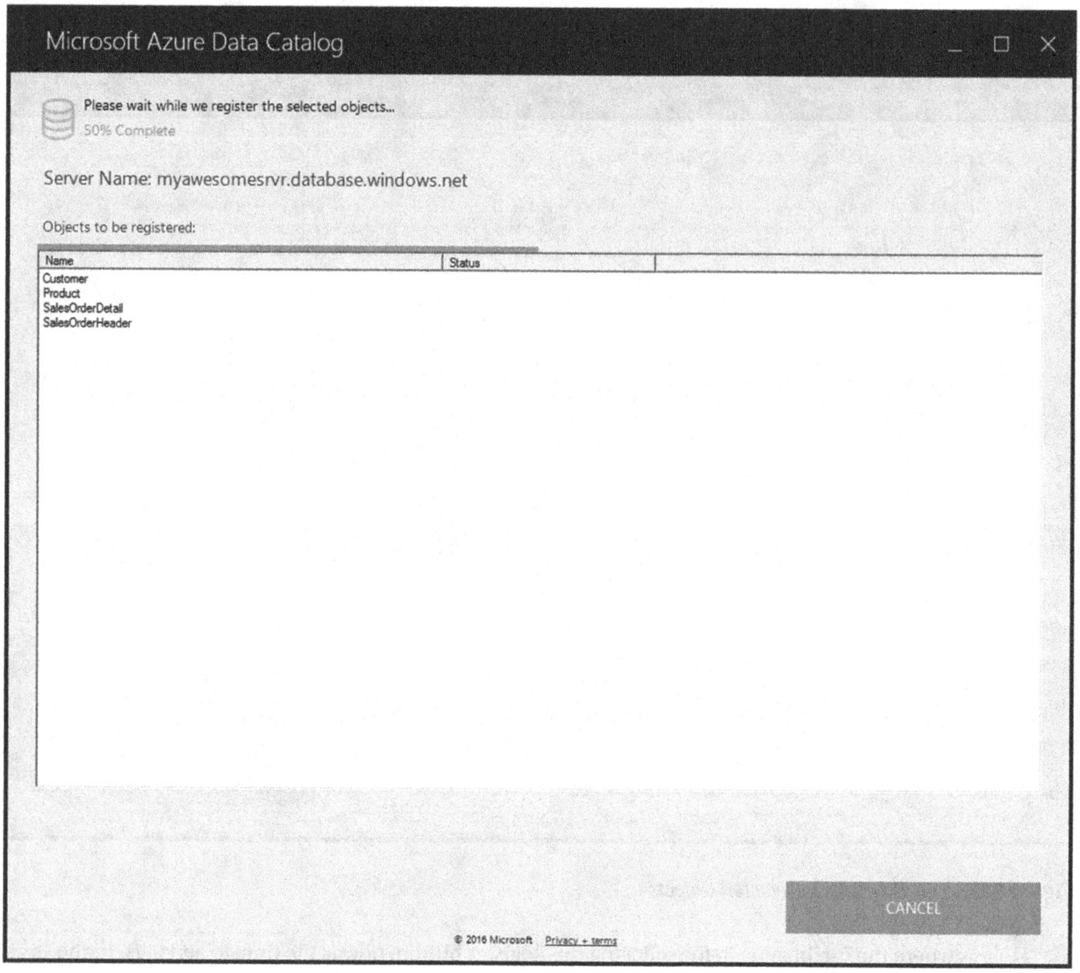

Figure 15-8. *Registering the objects*

As I mentioned earlier, your homework assignment is to go through the registration process again and this time select the Azure Data Lake data source and pick the file sitting the Data Lake Store. The registration process is similar so you should have no problems.

When the registration process is complete, you can register more objects (click this to go back and register the Azure Data Lake data source) or click View Portal (click this when you are done registering all the assets). Clicking the View Portal button opens up the Azure Data Catalog portal, displaying the registered assets, as shown in Figure 15-9.

Figure 15-9 shows the four tables from the Azure SQL Database, the AdventureWorks database itself, and the SensorData.csv file from Azure Data Lake Store.

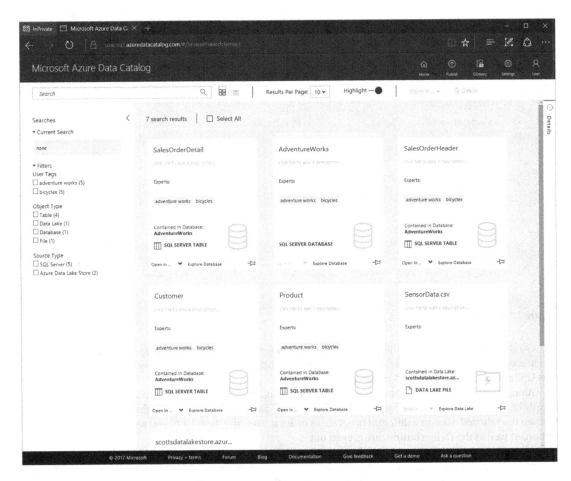

Figure 15-9. *Registered objects in Azure Data Catalog*

Before moving on, let's spend a few minutes on this page. Along the left are filters you can apply based on the registered assets. You can filter by tag, object type, and source type. Very cool. Along the top is the option of viewing the assets in grid view, as shown in Figure 15-9, or list view, as shown in Figure 15-10.

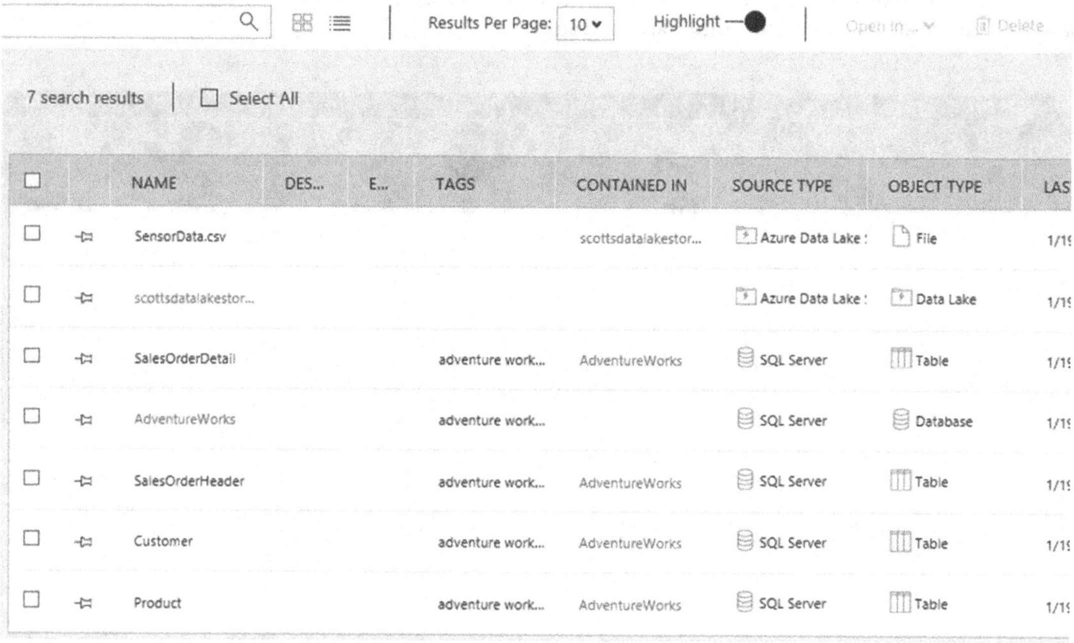

		NAME	DES...	E...	TAGS	CONTAINED IN	SOURCE TYPE	OBJECT TYPE	LAS
☐	⊣⊏	SensorData.csv				scottsdatalakestor...	Azure Data Lake :	File	1/19
☐	⊣⊏	scottsdatalakestor...					Azure Data Lake :	Data Lake	1/19
☐	⊣⊏	SalesOrderDetail			adventure work...	AdventureWorks	SQL Server	Table	1/19
☐	⊣⊏	AdventureWorks			adventure work...		SQL Server	Database	1/19
☐	⊣⊏	SalesOrderHeader			adventure work...	AdventureWorks	SQL Server	Table	1/19
☐	⊣⊏	Customer			adventure work...	AdventureWorks	SQL Server	Table	1/19
☐	⊣⊏	Product			adventure work...	AdventureWorks	SQL Server	Table	1/19

Figure 15-10. *Registered objects in grid view*

Go ahead and click the List View option. This view gives you an interesting perspective into the registered assets. You get a verbal description and icons, and the information is easier to view, such as the Last Updated and Last Registered columns.

Along the top, you will also notice the ability to turn highlighting on and off. You will see this in action shortly but I highly recommend you leave highlighting on. To the right of the highlighting option is the ability to open the selected asset in a different program in order to view the data. Until you select an asset, this option, as well as the Delete button, are greyed out.

Along the right of the screen you will see the Details blade. I will discuss this blade in more detail in the next section but this panel has a few tabs that display information regarding the object selected in the list.

One question that is frequently asked is about changing data sources and the impact this has on the registered data asset. For example, if a database table is altered to remove or add a column, are those changes reflected in Azure Data Catalog? The short answer is no. To update the metadata, the data source needs to be reregistered, at which point the metadata will be updated in the catalog and any annotations will be maintained.

OK, time to move on. With the assets registered, it is time to go discover these assets.

Discover Data Sources

Asset discovery is all about searching the catalog using search terms. The results will be the assets that have a match on any property in the asset. When the assets were registered, you applied tags to the assets and these tags are used to filter the search. You will see this momentarily but first let's explore the Details blade.

In the list of assets, click the Product asset. Automatically, the Details blade will slide out, showing the Properties tab. There are four tabs on the Details blade: Properties, Preview, Columns, and Documentation.

As shown in Figure 15-11, there is a lot you can do on the Properties tab, including adding a friendly name or a description of the asset, adding an expert (a subject matter expert for the selected asset, such as the owner of the data), and even adding additional tags. If you add a friendly name, this is the name that the user will see, so be careful how you use this field.

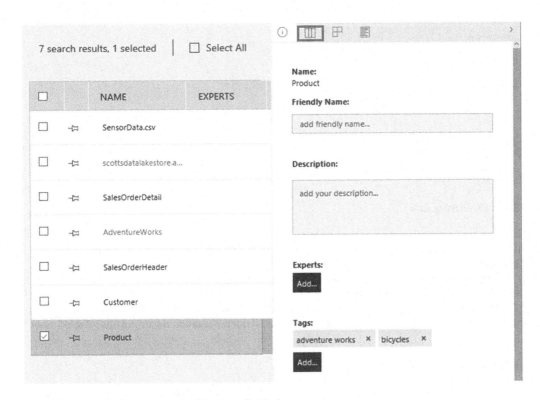

Figure 15-11. *The Properties tab of the Details blade*

As mentioned, this is where additional tags can be added. As users preview the data and discover the data source, they can add tags based on their needs for data discovery. Each user of Azure Data Catalog can add multiple tags for each asset, but only the user who created the tag can edit their own tags. Admins and asset owners can delete tags but not edit them.

Clicking the tab shows a preview of the data in the selected asset. As shown in Figure 15-12, the Preview tab shows a preview of the data in the Products table.

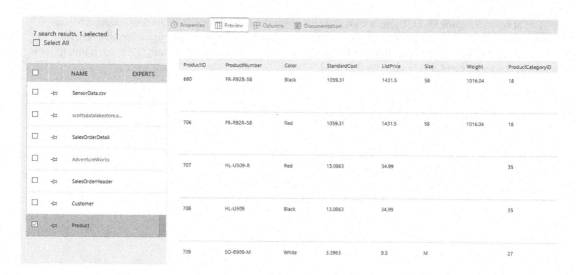

Figure 15-12. *Previewing data*

Going back for a moment, the preview of the data is available due to the fact that during the asset registration process you checked the Include Preview checkbox (see Figure 15-7). Remember also that checking that checkbox will copy up to 20 records from each table selected into the data catalog. Thus, it is recommended that you check this checkbox.

The Columns tab shows details about the columns in the table such as name and data type. The Documentation tab allows you to add a description and overall documentation regarding this selected asset.

Data Discovery

Let's talk about data discovery, which is accomplished via searches. In the search box, which is in the upper left, type in *bicycles*, as shown in Figure 15-13. Press Enter.

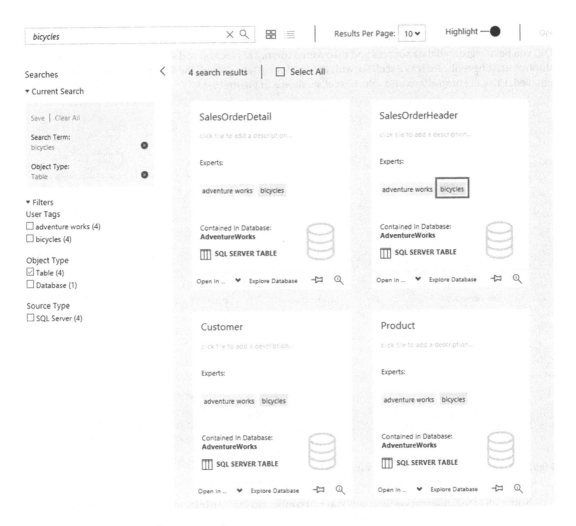

Figure 15-13. *Discovering data via search*

You will notice a couple of things, both of which are shown in Figure 15-13. First, because highlighting is turned on, the search keyword is highlighted in the search results. Second, only the assets that match the search keyword are returned. In this example, the four tables from the AdventureWorks database are returned and the keyword is highlighted.

This was a very simple search example, but there are other ways you can search. For example, you can discover assets via property scoping in the search. For example, you can type *tags:bicycles*. You can also do a Boolean search, such as *tags:bicycles AND objecttype:table*.

What is cool is that any and all executed searches can be saved. As you can see in Figure 15-13, in the Searches blade you can save a search by clicking the Save link. Again, very cool, especially when you have a more complicated search such as a query that has logical isolation via parentheses, such as *name:Customer AND (tags:bicycles AND objecttype:table)*.

Connect to Data Sources

OK, you have registered data sources and discovered them. This section will show how to connect to them through existing tools, such as Excel. So, with the Product asset selected, the Open In option will now be enabled. Click the drop-down and select Excel, as shown in Figure 15-14.

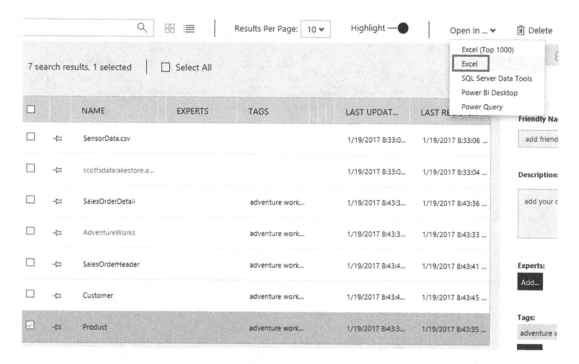

Figure 15-14. *Connecting to a Dataset via Excel*

Notice all of the different applications you can connect to that are integrated into Azure Data Catalog. It's a very nice list, including SQL Server Data Tools and Power BI.

OK, back to the example. Clicking on the Excel option will download a file called Product.odc. An .odc file is an Office Data Connection file and it provides the ability to connect to a data source via, in this case, Excel. Depending on the browser, you can either open it or save it (see Figure 15-15.) Open the file, which will open up Excel with a nice security notice. Click Enable on the notice.

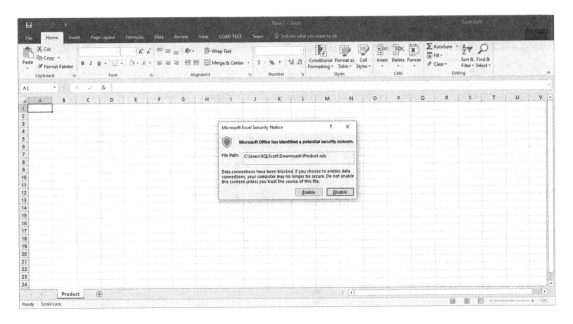

Figure 15-15. *Opening Excel*

Your next dialog will be the Import Data dialog, shown in Figure 15-16. Leave everything as is and click OK.

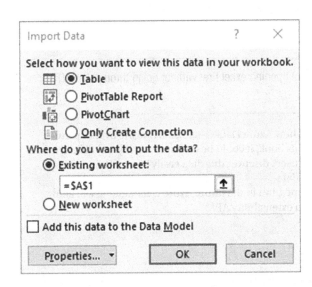

Figure 15-16. *Importing the data*

Since the data is coming from Azure SQL Database, you will be prompted to enter credentials to authenticate. Once authenticated, Excel will pull the data in and display it, as shown in Figure 15-17.

271

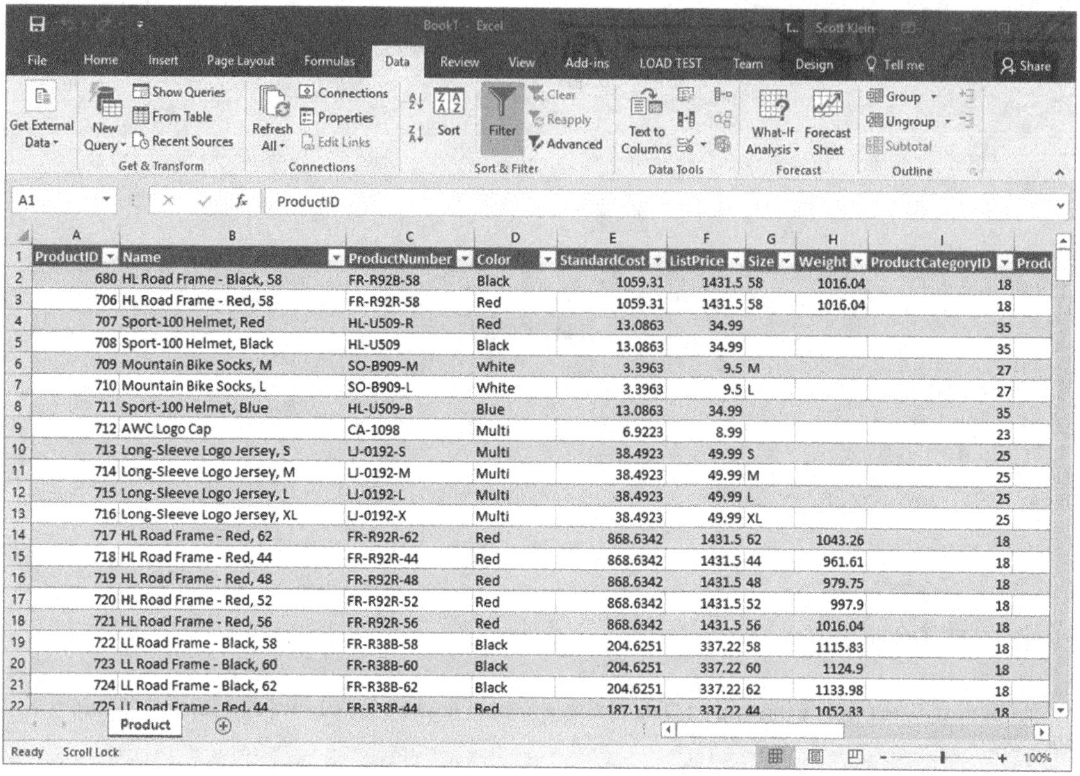

Figure 15-17. *Viewing the data in Excel*

■ **Note** Checking with engineering on one item about opening excel first without going through the ADC portal. Will update during AR.

Time for a quick review. Think about where and how Azure Data Catalog fits into the overall picture. Data is all over your organization, and in respect to this book, it could be coming from myriad different data sources including devices and sensors. How do your users discover this data easily? Azure Data Catalog makes it easy to register, discover, and consume that data.

Currently, it is not possible to add data source types, but in the future Azure Data Catalog will allow third parties to add new data source types through an extensibility API.

Summary

In this chapter, you learned about discovering data. Data is all over within your organization and it should be easily discoverable and understandable. You began by creating the Azure Data Catalog and then you registered different data assets from different data sources. You then discovered those data sources through searches and the different ways to search. Lastly, you looked at how to connect to the data sources and use different applications to consume the data.

CHAPTER 16

■ ■ ■

Azure Event Hubs

Chapter 3 covered Azure IoT Hub in detail, and as part of that chapter Azure Event Hubs was mentioned many times during the configuration of the Event Hub. Like IoT Hub, Event Hubs is an event processing service and it shares many similarities to Azure IoT Hub. However, there are some differences between the two services, which were briefly outlined in Chapter 3. One of the differences not discussed in Chapter 3 is the cost. Azure IoT Hub pricing is based on the total number of messages per day per unit. It is easy to calculate your cost. For example, 400,000 messages per day per unit will cost $50. Six million messages will cost $500.

Azure Event Hubs charges $0.028 per million events plus, depending on the pricing tier, charges for throughput units. However, Event Hubs also contains a throughput unit cost/hour. Event Hubs also offers a dedicated option, which is sold in capacity units for a fixed price. It's worth checking out.

However, don't let me or this comparison dissuade you. Event Hubs is still a viable solution and this chapter will provide a comparison between Azure IoT Hub and Azure Event Hubs so you can make your own decision.

Both are event processing and data streaming service and both can be used for real-time analytics. They are both excellent avenues for working with big data and IoT (Internet of Things) solutions. The decision becomes when to use which service and why.

This chapter will begin by providing an overview of Azure Event Hubs and then I will discuss in more detail the similarities and differences between Azure IoT Hub and Azure Event Hubs.

What Is Azure Event Hubs?

Much like Azure IoT Hub, Azure Event Hubs is a scalable event processing service and data streaming service with the ability to consume and process millions of events per second. Events can be received from event producers, including applications and devices, and routed for further processing. A key tenant of Event Hubs is the capability of high throughout and event processing and as such it does not implement many of the messaging abilities available through other Service Bus entities such as Topics.

Given Azure Event Hubs' event ingestion capability, low latency, and high reliability and scalability, there arises some confusion as to the difference between Azure Event Hubs and Azure IoT Hub, and in what scenario each service should be used. The following section will provide a comparison between the two services.

© Scott Klein 2017
S. Klein, *IoT Solutions in Microsoft's Azure IoT Suite*, DOI 10.1007/978-1-4842-2143-3_16

IoT Hub vs. Event Hubs Comparison

Table 16-1 outlines the major differences between Azure IoT Hub and Azure Event Hubs.

Table 16-1. Comparison Table

Area	IoT Hub	Event Hubs
Communication Patterns	Device-to-cloud and cloud-to-device	Device-to-cloud
Device State Information	Device twins can store and query device state information	No state information saved
Protocol Support	MQTT, MQTT via WebSockets, AMQP, AMQP over WebSockets, HTTP, and Azure IoT Protocol Gateway	AMQP, AMQP over WebSockets, and HTTP
Security	Per-device identity and revocable access control	Event Hubs-wide Shared Access Policies with limited revocation support.
Scale	Optimized to support millions of simultaneous connected devices	Up to 5,000 AMQP connections, limited number of simultaneous connections
Device SDKs	Support for large variety of platforms and languages	Supports .NET, Java, C, and AMQP and HTTP send interfaces
Operations Monitoring	IoT solutions can subscribe to device identity management and connectivity events to detect issues at the device level	Exposes only aggregate metrics
File upload	Solutions can send files from devices to cloud	Not supported
Message Routing	Up to ten additional endpoints with rules defining how messages are routed	Requires custom code for message dispatching

Again, both services enable event and telemetry input with high reliability and low latency, thus the decision of which to use will come down to the requirements of your solution and device support.

For example, how easy will it be to modify the example you built in Chapters 2 and 4 to send messages to Event Hubs? Can you send messages from a Raspberry Pi or Tessel to Event Hubs using the solutions created earlier? You will find out shortly. But first, an Event Hub needs to be created.

Creating an Event Hub

This section will walk through the process of creating an Azure Event Hub. Unlike Azure IoT Hub, the process requires multiple steps because an Event Hub is created at the namespace level. A namespace is used for addressing, isolation, and management, and all Event Hubs are created within the scope of the namespace.

Creating the Namespace

To get started, open up your favorite browser (if not already open) and navigate to the Azure portal at http://portal.azure.com/. Log in to the portal, click New, and then select Internet of Things ➤ Event Hubs, as shown in Figure 16-1.

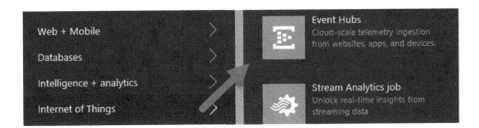

Figure 16-1. *Creating an Azure Event Hub*

Remember that the first step of creating an Event Hub requires the creation of a namespace, which is why the Create Namespace blade opens, as shown in Figure 16-2.

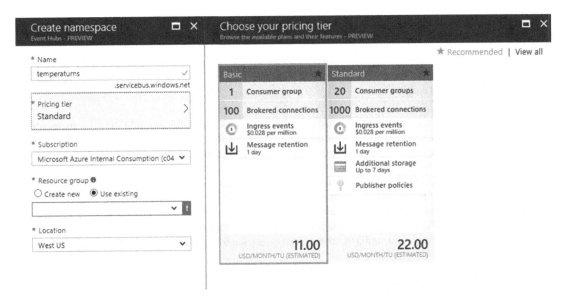

Figure 16-2. *Creating the namespace*

Provide a name for the namespace and then select the appropriate pricing tier. As of this writing, the two available options are the basic and standard tier. The standard tier offers more connections, additional storage, and publisher policies.

Select the appropriate resource group and location, and then ensure the "Pin to dashboard" option is checked. Once the Create Namespace blade is filled out and looks something like Figure 16-3, click Create. The creation only takes a moment and the Event Hubs tile will appear on your portal dashboard.

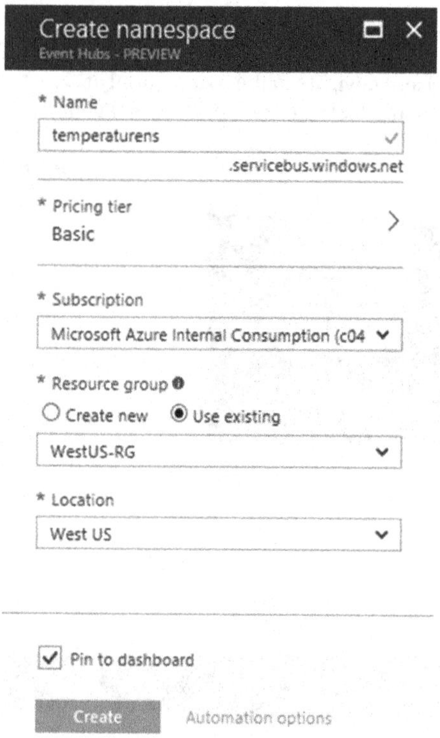

Figure 16-3. *Namespace configuration*

Before moving on, I'll provide a quick word on publisher policies. A publisher policy is security feature of Event Hubs that can be used to protect your processing pipeline. As you will see shortly, Event Hubs works with shared access policies, which in turn are used to create a SAS Token. For the purposes of this example, I selected the basic tier but in production you will want to consider the standard tier.

The next step is to create the actual Event Hub. The Event Hub is actually built around the managed classes within the Azure Service Bus namespace. The Service Bus is the component that provides the messaging infrastructure for relaying events and messages between devices and endpoints.

In the old version of the Azure portal, you first created the Service Bus and then you created the Event Hub. In the new Azure portal, you similarly create the namespace first and then create the Event Hub.

Creating the Event Hub

Once the namespace has been created, click the new Event Hubs tile on the dashboard, which will open the Properties and Overview blades, as shown in Figure 16-4. The options on the Properties blade are similar to other Azure services you have worked with throughout this book. Some differences to point out, which you will work with shortly, are the links for Shared Access Policies, Metrics monitoring, and Event Hubs.

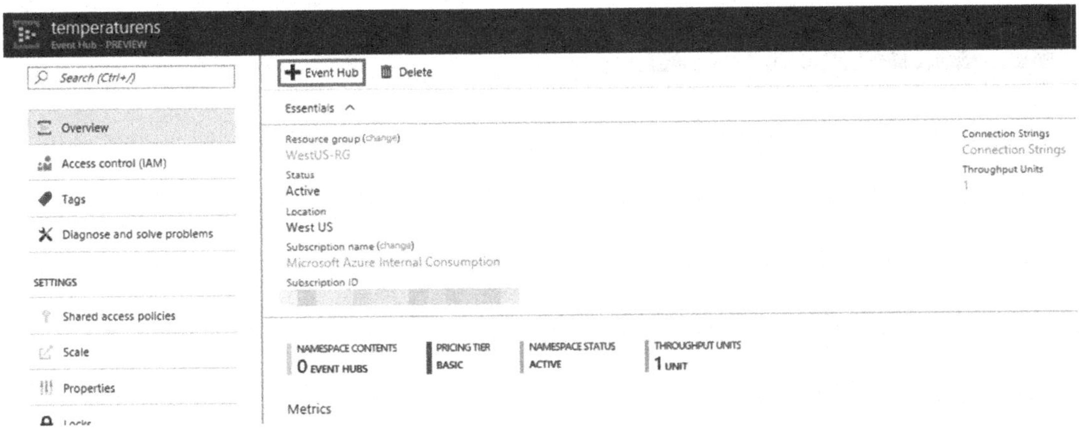

Figure 16-4. *Namespace Properties and Overview blades*

To create a new Event Hub, click the + Event Hub link at the top of the Overview blade, as highlighted in Figure 16-4. The Create Event Hub blade, shown in Figure 16-5, will open and, based on the pricing tier, will ask you to fill out certain information.

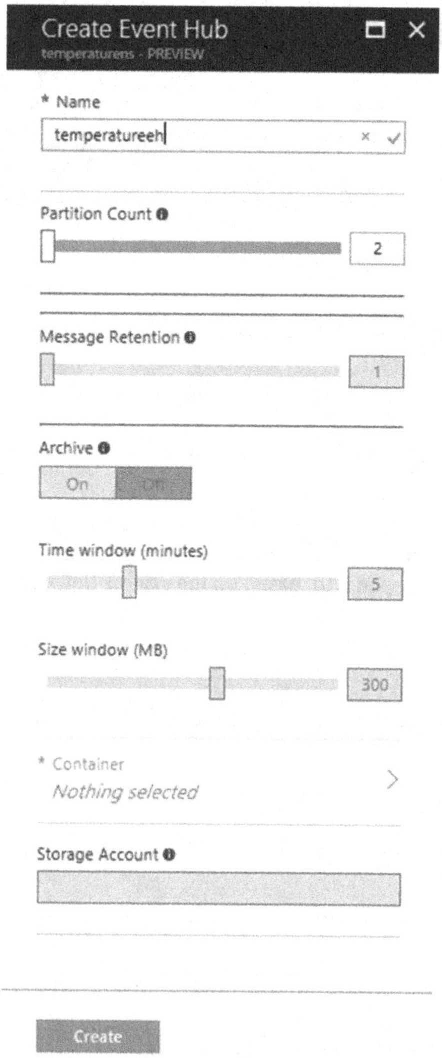

Figure 16-5. *The Create Event Hub blade*

Since this example selected the basic tier, the only options that need to be filled in are the name and the partition count. A partition is an ordered sequence of events that is held in an Event Hub. As new events arrive in the Event Hub, they are added to the end of the sequence.

Partitions cannot be changed, thus, like Azure IoT Hub, plan long term for scale when setting the partition count. For Azure Event Hub, the number of partitions correlates to the number of current readers you expect to have.

All partitions keep data for a configured amount of time (see Figure 16-2). You cannot explicitly remove messages from a partition; they will expire all on their own. Keep in mind also that each partition is independent of other partitions so they will grow at different rates.

Had you selected the standard tier, the other options on the blade would be enabled. You could have had the opportunity to add additional message retention time (in days) and to archive incoming messages to an Azure Blob storage account. By also sending incoming messages to a storage medium, you now have the ability to process the data in both real-time and batch mode on the same incoming stream of data, thus having both hot path and cold path capabilities on the same data.

You saw this simultaneous hot path/cold path scenario earlier in this book. As you learned in Chapter 6, Azure Stream Analytics provides the ability to send data to multiple hot path and cold path outputs including Power BI and Azure Blob Store or Azure Data Lake Store.

Click Create on the Create Event Hub blade to create the Event Hub (see Figure 16-5).

The creation of the Event Hub is quick but it is not ready to be used yet. The next step is to add the security component by defining the shared access policies.

Defining the Shared Access Policies

Event Hubs uses shared access signatures for Event Hub authentication. When creating a shared access policy, a SAS (shared access signature) token is produced using the name of the SAS key (policy) and is an encoded SHA hash of a URL. By using both the name of the key (policy) and the token, applications can be authenticated by Event Hubs.

To create a shared access policy, click the Event Hub link in the Properties pane, opening the Event Hub pane shown in Figure 16-6. The Event Hub pane will list all Event Hubs created within the selected namespace.

NAME	STATUS	MESSAGE RETENTION	PARTITION COUNT
temperatureeh	Active	1	2

Figure 16-6. *Event Hubs within the selected namespace*

Within the Event Hub pane, click the Event Hub created above, thus opening the Event Hub Properties and Overview panes shown in Figure 16-7.

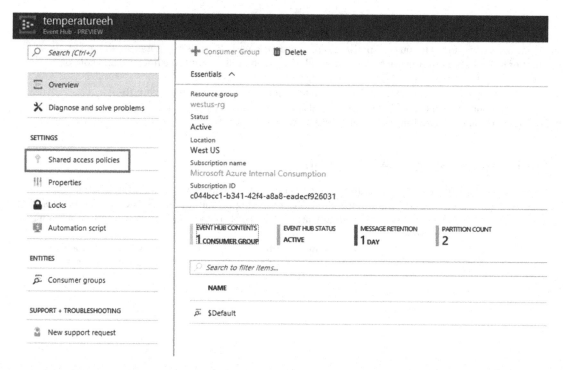

Figure 16-7. *Event Hub Properties and Overview blades*

Within the Properties pane for the Event Hub, click the "Shared access policies" link, highlighted in Figure 16-7. The Shared Access Policies pane will appear, listing all created policies. All new namespaces are created with a default root policy called RootManageSharedAccessKey, and this policy has all permissions. Since you don't log on as "root," there is no need to use this policy.

Click the +Add link at the top of the Shared Access Policies blade, opening the Add New Shared Access Policy blade, shown in Figure 16-8.

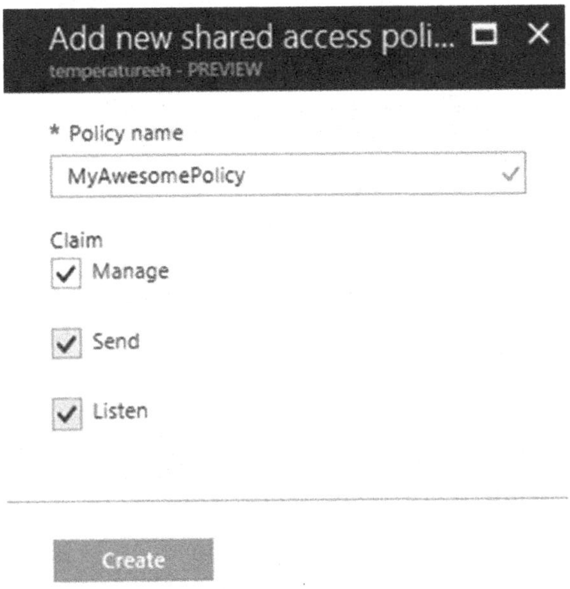

Figure 16-8. *Creating a shared access policy*

In the Add New Shared Access Policy blade, enter a policy name and select the permissions this policy will have. Selecting Manage will automatically select Send and Listen. For the purposes of this example, select Manage.

Click Create, at which point the policy will be created and a primary key and secondary key will be created and assigned to the policy. You will then be taken back to the Shared Access Policies pane listing the RootManageSharedAccessKey policy and the policy you just created.

Click the policy you just created, opening the blade containing the details of the shared access policy, shown in Figure 16-9. On this blade, you also have the ability to modify the permissions of this policy. Also listed on this blade are the primary key and secondary key as well as a primary and secondary connection string. These connection strings will be used in applications needing to connect and send data to Event Hubs.

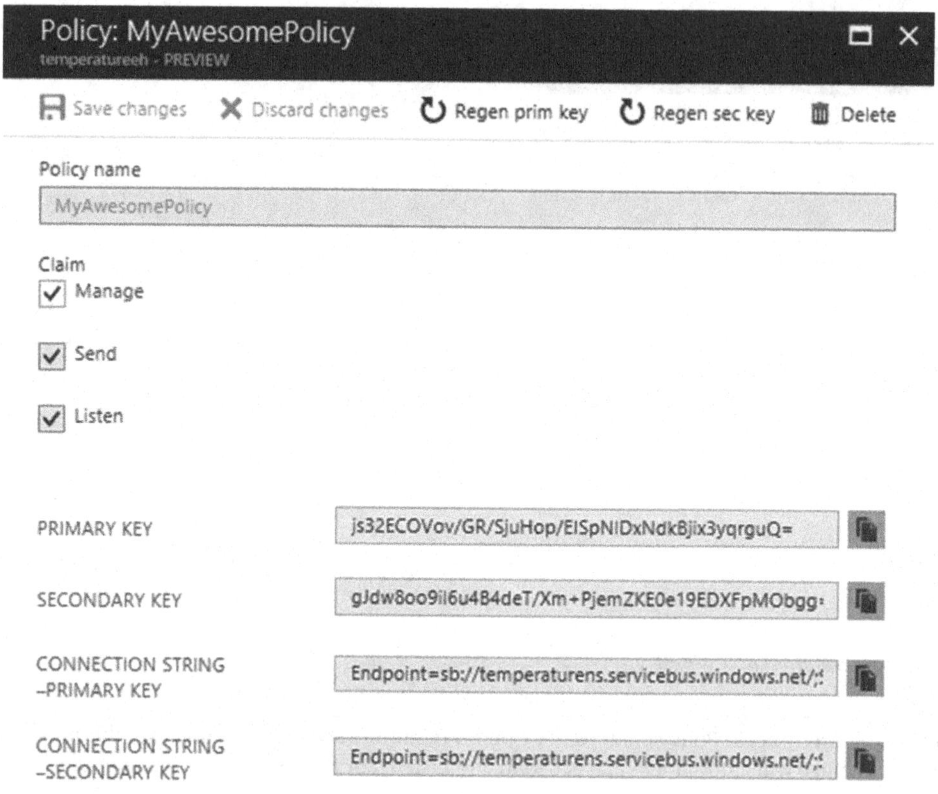

Policy: MyAwesomePolicy
temperatureeh - PREVIEW

🔲 Save changes   ✖ Discard changes   ↻ Regen prim key   ↻ Regen sec key   🗑 Delete

Policy name

MyAwesomePolicy

Claim
☑ Manage

☑ Send

☑ Listen

PRIMARY KEY   js32ECOVov/GR/SjuHop/EISpNlDxNdkBjix3yqrguQ=

SECONDARY KEY   gJdw8oo9il6u4B4deT/Xm+PjemZKE0e19EDXFpMObgg=

CONNECTION STRING   Endpoint=sb://temperaturens.servicebus.windows.net/;‹
–PRIMARY KEY

CONNECTION STRING   Endpoint=sb://temperaturens.servicebus.windows.net/;‹
–SECONDARY KEY

Figure 16-9. Details of the shared access policy

Shared access policy keys can be regenerated at any time but any previous shared access policy keys will be invalidated. Also keep in mind that you can have only 12 policies attached to a single namespace. It is worth mentioning here that this models the Service Bus permission sets and finish.

At this point, the Event Hub is created and configured and ready to be put to use. Before moving on, click the Copy button next to the primary connection string because you will need that shortly.

Sending Messages to the Event Hubs

This section will create a very simple application to send simple messages to the Event Hub created above. Create a new C# Console Application in Visual Studio, as shown in Figure 16-10. Give the solution a meaningful name and then click OK to create the solution.

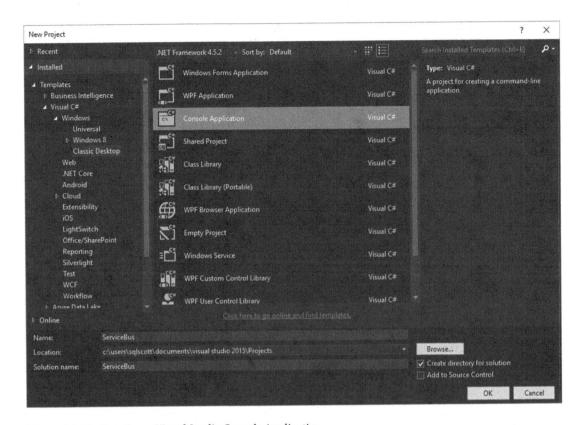

Figure 16-10. *Creating a Visual Studio Console Application*

When the solution loads, select the Tools menu and then select Nuget Package Manager ➤ Package Manager Console. In the Package Manager Console window, type in `Install-Package WindowsAzure.ServiceBus`, as shown in Figure 16-11, and then press Enter.

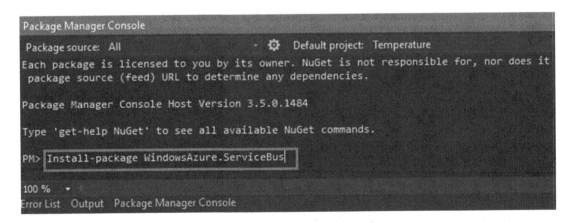

Figure 16-11. *Installing the Microsoft Azure Service Bus Nuget Package*

This Microsoft Azure Service Bus package includes all the libraries necessary for working with service bus queues, topics, Event Hubs, and relay operations. The current version of this package is 3.4.4. Visit www.nuget.org/packages/WindowsAzure.ServiceBus to learn more about this package.

Once the package is installed, open the Program.cs file and add the following using statements below the existing using statements:

```
using System.Threading;
using Microsoft.ServiceBus.Messaging;
```

Next, add the following two lines to the Program class above the Main() method:

```
static string eventHubName = "temperatureeh";
static string connectionString = "<connectionstring>";
```

Next, add the following line of code inside the Main() method:

```
SendMessages();
```

The next step is to create the SendMessages() method, so add the following code below the Main() method:

```
static void SendMessages()
{
    var eventHubClient = EventHubClient.CreateFromConnectionString(connectionString,
eventHubName);
    while (true)
    {
        try
        {
            var message = "Saying Hello at :" + DateTime.Now.ToString();
            Console.WriteLine("{0}", message);
            eventHubClient.Send(new EventData(Encoding.UTF8.GetBytes(message)));
        }
        catch (Exception exception)
        {
            Console.ForegroundColor = ConsoleColor.Red;
            Console.WriteLine("{0} > Exception: {1}", DateTime.Now, exception.Message);
            Console.ResetColor();
        }

        Thread.Sleep(200);
    }
}
```

You are not quite done yet. You still need to update the eventHubName and connectionString variables. I named my Event Hub temperatureeh, so you will need to change yours if you named it differently. You can find the connection string back in Figure 16-9. At the time I had you copy the primary connection string. If you still have it, replace the <connectionstring> with the actual connection string.

Now, I'm going to save you a bit of troubleshooting here. At the end of the connection string is an EntityPath parameter. You need to remove it; otherwise you will get an error when you run the solution stating that the "EventHub name should not be specified as EntityPath while using this Overload." So, remove the `EntityPath` parameter and value as well as the preceding semicolon. For example, my connection string included the following `EntityPath` parameter, which I removed:

`;EntityPath=temperatureeh`

Once removed, you are good to go! Build the solution to ensure no compile errors and then run the solution. A command window will run, showing the events being sent to the Event Hub, shown in Figure 16-12.

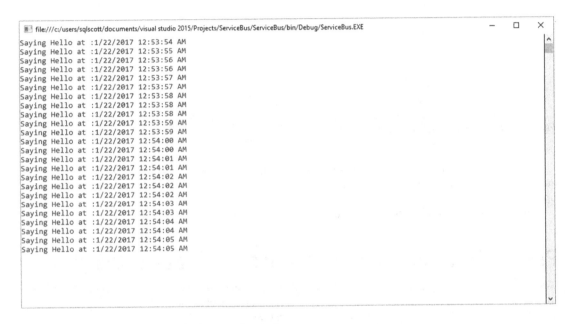

Figure 16-12. *Displaying events in the console window*

The messages are appearing in the console window, but how do you know they are being received by the Event Hub? Back in the portal, close all the open blades and navigate back to the namespace Properties and Overview blades.

In the Properties blade, click the Metrics link to open the Metrics blade. The Metrics blade provides real-time insight into what is happening within the namespace. On the left is a list of available metrics to show in the graph. There is a significant list of metrics to choose from, include incoming and outgoing messages, errors, and more.

By default, no metrics are selected, so select the Incoming Messages option which will, by default, plot a line graph for the past hour of the incoming messages. The line graph updates roughly every 30-60 seconds so the graph may not update immediately. Trust me, though, the graph will update soon enough and the line will move. Figure 16-13 shows the count of messages my application sent when I ran it.

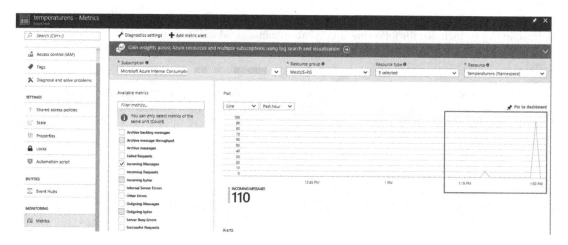

Figure 16-13. *Viewing the metrics in the portal*

You can change the type of chart plot from line to bar (line looks cooler in this case, in my opinion) and you can change the chart timeframe. The minimum is Past Hour, the maximum is Past Week. The other option is Today.

Are you done? Not quite, because I'd like to point out one more thing.

Pulling Messages from Event Hubs with Stream Analytics

Now that messages are coming into Azure Event Hubs, you can act on them just like you did with Azure IoT Hub in earlier chapters. Like IoT Hub, Azure Event Hubs can collect millions of events, allowing Azure Stream Analytics to pick up and process the events in real time.

As a quick example, close out of the Event Hub blades in the Azure portal and go back to the main dashboard. On the dashboard, click the Stream Analytics tile to open the Stream Analytics job created in Chapters 6 and 7. In the Overview pane, click the Inputs tile to display the Inputs blade. In the Inputs blade, click the +New link to open the New Input blade.

In the New Input blade, click the Source drop-down and select the Event Hub option. For the Service bus namespace, select temperatures, and then make sure the correct event hub name and event hub policy name are selected, as shown in Figure 16-14.

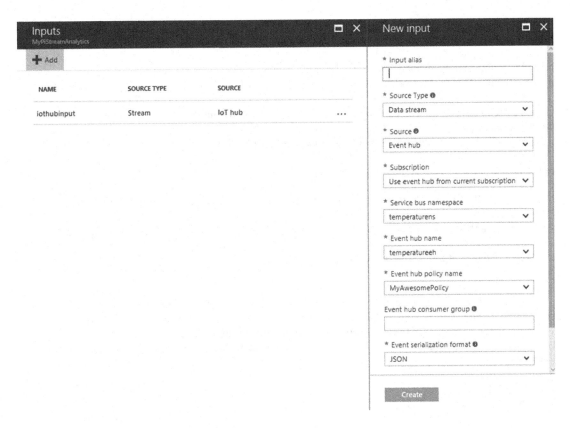

Figure 16-14. *Creating new Stream Analytics Event Hub input*

Don't click Create because first, you probably haven't given it a name yet, and more importantly, you don't have any temperature sensor data coming in yet. So, go ahead and give it a name such as ehinput and then click Create.

Now comes the exciting part, which is to modify the original Visual Studio solution created in Chapter 2. In Chapter 4, the solution was modified to send temperature data to Azure IoT Hub, but now you'll modify it again to send the data to Azure Event Hubs instead.

In order to work with Azure Event Hubs, you need to install the library that will allow you to do so. It's available on Nuget here:

```
www.nuget.org/packages/Microsoft.Azure.EventHubs/
```

This package has recently been updated to support UWP apps, so the next step is to modify the temperature application created earlier. Open Visual Studio, open the project from Chapter 4, and install the Azure Event Hubs package. You have done this enough times throughout this book so I won't walk you through it again.

Once it is installed, you will notice a new reference in Solution Explorer. You need to add a using statement to reference it:

```
using Microsoft.Azure.EventHubs;
```

Next, copy the eventHubName and connectionString constants from the earlier example and paste them below the IOTHUBCONNECTIONSTRING constant:

```
private const string eventHubName = "temperatureeh";
private const string connectionString = "<connectionstring>";
```

As stated in the earlier example, be sure the EntityPath parameter is removed from the end of the connection string.

Next, add the following method below the readSensor() method:

```
private static async Task MainAsync(string ehMessage)
{
    var eventHubClient = EventHubClient.CreateFromConnectionString(connectionString);

    await eventHubClient.SendAsync(new EventData(Encoding.UTF8.GetBytes(ehMessage)));

    await eventHubClient.CloseAsync();
}
```

The MainAsync method essentially does the same thing the IoT Hub code does in that it creates a connection to the Event Hub created above and then sends the event message to the Event Hub.

Next, in the readSensor() method, comment out the three lines of code that send the message to IoT Hub and add the line of code that sends the message to the Event Hub. The code should now look like the following:

```
//send to listbox
listBox.Items.Add(json.ToString());

//send to azure IoT Hub
//Message eventMessage = new Message(Encoding.UTF8.GetBytes(json));
//DeviceClient deviceClient = DeviceClient.CreateFromConnectionString
(IOTHUBCONNECTIONSTRING);
//await deviceClient.SendEventAsync(eventMessage);

//send to Azure Event Hubs
await MainAsync(json.ToString());
```

Compile the solution to make sure all is well, and just like you did in Chapter 4, run the solution by clicking the Remote Machine button on the toolbar. Pretty soon you will be seeing the JSON appear in the ListBox and more importantly, you'll be able to see the messages coming into Event Hubs, as you saw in the previous example by viewing the Metrics in the Azure portal in Figure 16-13. At this point you have event messages going into Event Hubs, but they are not being routed anywhere. Your homework assignment is to modify the Azure Stream Analytics query to pull data from the new Event Hub input and send the data to an output. You can choose what that output is.

That was quite the look at Azure Event Hubs, and one that should give you insight into how it compares to Azure IoT Hub. Now, go forth and IoT!

Summary

Azure Event Hubs shares many similarities with Azure IoT Hub, but there are also some significant differences. In this chapter, you were introduced to Azure Event Hubs and then you spent some important time learning the key differences between the two event processing services.

With that foundation built, you then walked through creating an Azure Event Hubs namespace and Event Hub. The importance of security in Event Hubs via shared access policies was discussed and then applied by building a shared access policy within the created Event Hub.

Lastly, you walked through the creation of an application, which sent messages to the Event Hub. This was verified by checking the metrics within the Event Hub in the Azure portal and seeing the number of messages being received.

Index

Get the eBook for only $5!

Why limit yourself?

With most of our titles available in both PDF and ePUB format, you can access your content wherever and however you wish—on your PC, phone, tablet, or reader.

Since you've purchased this print book, we are happy to offer you the eBook for just $5.

To learn more, go to http://www.apress.com/companion or contact support@apress.com.

Apress®

All Apress eBooks are subject to copyright. All rights are reserved by the Publisher, whether the whole or part of the material is concerned, specifically the rights of translation, reprinting, reuse of illustrations, recitation, broadcasting, reproduction on microfilms or in any other physical way, and transmission or information storage and retrieval, electronic adaptation, computer software, or by similar or dissimilar methodology now known or hereafter developed. Exempted from this legal reservation are brief excerpts in connection with reviews or scholarly analysis or material supplied specifically for the purpose of being entered and executed on a computer system, for exclusive use by the purchaser of the work. Duplication of this publication or parts thereof is permitted only under the provisions of the Copyright Law of the Publisher's location, in its current version, and permission for use must always be obtained from Springer. Permissions for use may be obtained through RightsLink at the Copyright Clearance Center. Violations are liable to prosecution under the respective Copyright Law.

Printed in the United States
By Bookmasters